T0073777

CANADA AND CLIMATE CHANGE

CANADIAN ESSENTIALS
Series editor: Daniel Béland

Provocative thinking and accessible writing are more necessary than ever to illuminate Canadian society and to understand the opportunities and challenges that Canada faces. A joint venture between McGill-Queen's University Press and the McGill Institute for the Study of Canada, this series arms politically active readers with the understanding necessary for engaging in – and improving – public debate on the fundamental issues that have shaped our nation. Offering diverse and multidisciplinary perspectives on the leading subjects Canadians care about, Canadian Essentials seeks to make foundational and cutting-edge knowledge more accessible to informed citizens, practitioners, and students. Each title in this series aims to bolster individual action in order to support a better, more inclusive and dynamic country. Canadian Essentials welcomes proposals for concise and well-written books dealing with far-reaching and timely Canadian topics from a broad swath of authors, both within and outside of academia.

Canada and Climate Change

WILLIAM LEISS

Published for the McGill
Institute for the Study of Canada
by
McGill-Queen's University Press
Montreal & Kingston • London • Chicago

ISBN 978-0-2280-0916-0 (cloth)
ISBN 978-0-2280-0985-6 (ePDF)
ISBN 978-0-2280-0986-3 (ePUB)

Legal deposit fourth quarter 2022
Bibliothèque nationale du Québec

Printed in Canada on acid-free paper that is 100% ancient forest free
(100% post-consumer recycled), processed chlorine free

Funded by the Government of Canada   Financé par le gouvernement du Canada

Canada Council for the Arts   Conseil des arts du Canada

We acknowledge the support of the Canada Council for the Arts.
Nous remercions le Conseil des arts du Canada de son soutien.

Library and Archives Canada Cataloguing in Publication

Title: Canada and climate change / William Leiss.
Names: Leiss, William, 1939– author.
Description: Series statement: Canadian essentials ; 2 | Includes bibliographical
  references and index.
Identifiers: Canadiana (print) 20220395160 | Canadiana (ebook) 20220395292 | ISBN
  9780228009160 (hardcover) | ISBN 9780228009856 (PDF) | ISBN 9780228009863
  (ePUB)
Subjects: LCSH: Climatic changes—Canada. | LCSH: Climatic changes—
  Government policy—Canada. | LCSH: Environmental policy—Canada.
Classification: LCC QC903.2.C3 L45 2022 | DDC 363.738/740971—dc23

This book was typeset in Minion Pro.

For my wife, Heidemarie

Contents

Tables and Figures

Preface

Presenting a short new book on the issue of climate change might well be regarded as a monumentally foolish endeavour. Among other things, there exists already enough written material on this subject to require a lifetime of study. Moreover, the matter has elicited such bitter controversy over both its intellectual and practical aspects that for many fossil-fuel-producing countries, including Canada, these controversies have paralyzed the public-policy decision-making process for a long time. On a practical level, all major nations of the world have struggled for thirty years to come up with a credible response to this issue and, as of the time of writing, they are still doing so. The most common response of politicians to the policy dilemmas they face is to make earnest commitments for reducing their greenhouse-gas emissions, with delivery dates so far into the future that they will certainly no longer be in office – and quite possibly will have passed on to their ultimate just rewards – by the time the bill comes due. Little has changed in the past thirty years.

Nevertheless, this book presupposes that the need still exists for further guidance in the matter of climate change, including a focus on Canada in particular. Given the scope of the issue, perhaps only a volume written with a specific audience in mind, and a strategy for presenting those readers with a specific approach, might hope to offer an incremental benefit at this time. At the same time, the timing might be fortuitous. The federal government now in office and at least two major provincial governments, British Columbia and Quebec, have announced newer and firmer commitments for addressing Canada's responses in terms of public policies. The federal partner even enshrined its commitments in legislation in 2021. So perhaps the time is ripe.

The intended audience for this book is the educated general reader rather than academic specialists, who might well opine that they have little need of it. Second, the author hopes that it might usefully be assigned to secondary-school students, as well as college and university students enrolled in introductory courses. For this reason, there is a much-reduced apparatus for source citations, although the references section provides ample chapter-by-chapter suggestions for pursuing further study. This guidance includes the huge resources available on the World Wide Web, but with explicit cautions about how to exploit these resources with due care.

The strategy for the discussion may appear odd at first, especially in the focus it adopts in the early chapters, which are devoted to summarizing in clear language the scientific account of climate change. Since the author does not hold any academic credentials in this area, this choice may be regarded as at least as foolish as the one mentioned above. However, it is based on a simple educational calculation: the ordinary citizens of Canada and elsewhere, most of whom are not (like the author) expert in any of the disciplines employed in the field of climate science, need to have a firm grasp of *how and why* the scientific community has approached the problem of climate change in the way it has. Specifically, we citizens need to be able to answer this straightforward question: What is climate and how does it differ from weather? Simply stated, if we have not all begun with adequate knowledge on this point firmly in hand, there is no point in advancing any further in trying to form a judgment on the issue of climate change.

The sources for the discussions throughout the book have been carefully selected and evaluated as to their credibility and reliability. Not everyone will agree with the merits of the selection, but so be it. This review of the scientific study of climate includes illustrative examples drawn from its approach to the earth's deep past, beginning billions of years ago, as well as the types of evidence on which its findings are based. I have also looked into the highly charged area of how climate scientists make their estimates about *the likely course of the climate in the near and distant future*, under various scenarios, and why they think that their findings should be regarded by the rest of us as being reliable.

The approach in the later chapters then reviews the major current issues in the area of climate change that have a bearing on the kinds of choices that citizens in Canada and elsewhere are facing now and will be facing in more intensive form in the coming years. These major issues are:

1   Mitigation. How and why Canada should reduce its greenhouse-gas emissions and how rapidly and intensively must it do so?

2   Adaptation. What is likely to happen to the Canadian population in terms of impacts in the far North, on agriculture, on the coasts (rising sea levels), etc.?

3   Equity. What responsibility do Canadians have to achieve certain goals in the context of the global population, where serious inequalities exist in terms of climate change?

This is the main point to be made through the chapters to follow: *Once the reader has achieved the first stages of understanding – what climate and climate change is and with what methods scientists have sought to describe it – the rest will be fairly easy.* To facilitate the reader's knowledge about climate science and its methods, I have provided throughout the chapters many examples of the evidence scientists have assembled in order to form a clear picture of the many different stages and events in the earth's climate history. The demand to show credible and convincing evidence underlies every causative proposition in the work of climate scientists.

This book seeks to recount in summary fashion two parallel stories. The first tells of the progress of climate science and its core contention, namely, that an unrestricted growth in human-caused greenhouse-gas emissions constitutes "dangerous interference" with the climate system. The second is an account of how the nations of the world have responded to the core contention of climate science. What links the two stories is a recognition of the need for all nations collectively to stop the rise of greenhouse gas (GHG) concentrations in the atmosphere. What makes the story relevant to the present day is that so far these nations – including Canada – have failed to respond adequately to that need.

Learning from scientists about the broad evidentiary basis for the understanding of climate and climate change then makes it crystal clear what we as citizens should and must do. We will realize that we must control and eventually eliminate our human-originated GHG emissions; that we must be ready to adjust to the climate-change impacts on the horizon; and that we must help the still-developing nations of the world to acquire the requisite alternative energy-production technologies that will enable them to avoid creating GHG emissions. These commitments will not be cost-free, but I believe Canadian citizens will come to agree, once they have understood climate change, that they have a duty to support and fund all three of those objectives with their taxes – and in all respects these citizens will become convinced over time that they are therefore doing the right and proper thing.

CANADA AND CLIMATE CHANGE

Introduction

Humanity has faced many truly catastrophic risks in the past, both long ago and in recent times, but it has never faced a risk such as climate change. By the term "catastrophic risks," I refer to experiences of terrible loss of life as well as economic disaster. To take only one example from the distant past, the bubonic plague, beginning in the fourteenth century, killed one-third of the entire population of Europe. In the twentieth century we faced two world wars, involving massive destruction and tens of millions of casualties. After that, the Cold War, ruled by the doctrine of "mutually assured destruction," involved the threat of large-scale nuclear warfare, initiated either deliberately or accidentally, which quite possibly would have ended civilization in the most highly developed societies on the globe. In addition, of course, there were the pandemics of 1918–20 and 2020–22.

Most of the catastrophic risks we have faced to date have had one thing in common: the disastrous consequences stemming from them appeared suddenly in the human environment (or would have done, had large-scale nuclear warfare occurred). In every case mentioned here, and many others, people everywhere felt the adverse impacts at once; moreover, in the case of all the modern risks, the causes were apparent to the naked eye. There was no good reason for confusion about what had happened and why during the two infectious-disease pandemics or the two world wars, or about what would have followed upon the instantaneous obliteration of whole cities had ballistic missiles been launched to deliver their payloads of hydrogen bombs.

When we say that something is a risk, we mean that we have identified in advance the causes and consequences of a possible event that might entail some harm to us. To say it "might" happen means that there is some estimated chance or *likelihood* that it is capable of causing harm, at some time in the future, either near or far. When we know enough about it, we can even try, sometimes with a high degree of confidence, to estimate what might happen. By now we are all familiar with the daily weather forecasts that tell us, for example, that there is a 40 per cent chance of rain or snow on some upcoming day. Also, when serious risks arise, citizens need to know how to interpret warnings given in terms of probability in order to protect themselves and their families. During the earlier stages of the public-health crisis brought on by the COVID-19 pandemic, officials reported that the newer variants of the virus were 60 per cent more infectious than earlier ones, and that there was a chance that a third wave of the pandemic would occur; this prediction turned out to be correct. This type of expert judgment can help us to plan measures in advance to reduce the threat. However, in order to make such planning effective, citizens must understand the nature of the risk and continue to take the necessary actions during the pandemic, such as mask-wearing, social-distancing, lockdowns, travel restrictions, and others.

We often do look back, after something bad has happened to us, and wonder if we might have foreseen it and, having done so, taken precautions to avoid it or lessen its damages. To take an earlier instance, by at least the year 1912 many observers were convinced that a war involving many nations on the European continent was almost certainly inevitable, and in 1914 it happened. The probability of another and similar war was likewise thought to be very high by 1938 at the latest, and it began in September 1939. A pandemic specifically caused by a viral pathogen of the influenza or coronavirus family was thought to be highly likely to occur for some years before it actually appeared in December 2019. Many Western governments had done what they called "pandemic planning" following the SARS episode in 2003, but by 2019 they had grown complacent and had dismantled their early-warning units, and so they were woefully unprepared when COVID-19 struck. But here is the important point: *No one*, neither expert nor members of the public – unless wilfully blind or misled – could fail to have recognized the disaster for what it was once it began to unfold, nor could they not realize, quite quickly after its onset, what terrible consequences would likely ensue.

But never before has humanity faced a risk such as climate change. As compared with the risks of pandemics and wars, economic collapse, and powerful natural disasters such as hurricanes, where harms and costs quickly become so readily apparent, the potential harms associated with climate change cannot easily be grasped. Moreover, the worst impacts are predicted to evolve gradually over the course of decades and centuries and will not be apparent to the naked eye for many years yet to come. Other crises having immediate and severe consequences, whether pandemics or economic turmoil or military tensions among powerful nations, have the capacity to distract both politicians and citizens from paying attention to the risk of climate change. We can see this with the war in Ukraine in 2022.

As already noted, the perspective I adopt for the discussion to follow is based on the risk approach. Risk is "the chance of harm." Every one of us confronts a myriad of risks every day, beginning with those that will have an immediate harmful impact on our well-being: automobile accidents, dangerous drugs, food contamination, exposure to pathogens, and many others. Others have potentially harmful effects, which materialize only over long periods of time, such as cancers caused by smoking and heart disease caused by high cholesterol or many other factors. Some, however, like the impacts of global warming are multi-generational risks, a term by which we refer to current trends that will reveal their worst consequences only in the lifetimes of our children and grandchildren. (See appendix 2 for the latest summary of climate-change risks.)

It is important for the reader to know upfront that in this volume I will not deal with the so-called debate about whether or not anthropogenic (human-caused) climate change is "real," as opposed to something else, including the alleged possibility that it might simply be a hoax perpetrated by scientists hoping for additional funding for their research programs. Since I have no scientific expertise in the matter, there would be little point in my trying to figure out whether one or more climate-change "deniers" might be correct. In this regard I am in the same position as most other Canadians, who might ask themselves, "How should I decide what or whom to believe about climate change?" But this is a question which most of us have to ask ourselves about every risk that threatens our well-being, ranging from personal issues about vaccination for infectious diseases, or our young children's experimentation with psychoactive drugs such as alcohol and cannabis, to large social issues such as global warming. I will return to this theme in chapter 4.

Every specific risk I have mentioned so far, as well as dozens of others, has a detailed and complex scientific analysis that characterizes it. This analysis starts with expertise in chemistry, physics, and biology, moves on to applied fields such as medicine or engineering, and ends up with statistics and epidemiology. A typical end point in this process is the following statement from the website of the US Centers for Disease Control and Prevention (CDC): "People who smoke cigarettes are 15 to 30 times more likely to get lung cancer or die from lung cancer than people who do not smoke." Very few of us are familiar with the huge trove of biological, medical, and statistical knowledge that lies behind an apparently straightforward sentence such as this one. Anyone who is considering taking up regular smoking or vaping, whether of tobacco or marijuana or both, is free to ask herself: "Should I believe this statement?" "Could it be a hoax?"

In search of an answer, someone can do what his or her parents could not have done at a similar age: namely, undertake a quick and easy Internet search. But these days, having entered "smoking and lung cancer," one would find dozens of pages of entries explaining why this is a "real" risk without running across any plausible denials. (Had earlier generations been able to do such a search, they would have found many websites, sponsored by tobacco companies, which would have asserted that the association between smoking and lung cancer was unproven.) In the present day, if one puts the question "Is global warming real?" into one's search engine, as I did in December 2020, he or she would have found on the very first page of results both a site which reports that "31,000 scientists" say that there is "no convincing evidence" that humans cause global warming *and* another that states, "97% [of climate scientists] think human activity is a significant contributing factor in changing mean global temperatures." One would also find on the very first page of results a long, informative, clearly written, and authoritative Wikipedia entry under the title "global warming controversy." But if one is already somewhat persuaded that global warming might be a hoax, a dedicated search to confirm this suspicion will readily turn up a rich trove of ostensibly supportive arguments.

In the fully formed social-media environment in which we now live, stirring controversies on just about any topic, from vaccination safety to US presidential election results, are the powerful drivers of individual attention and the formation of beliefs. It is therefore unsurprising that global warming should be the subject of some sustained ideological warfare. To be fair, "manufactured" controversies over risk assessments were being stirred up long before the Internet age. The tobacco industry undertook a decades-long legal battle and disinformation campaign on the issue of smoking and lung cancer, one that became so prominent that it was the subject of the 1999 Hollywood movie *The Insider*.

The asbestos industry also launched a long campaign to deny the health risks of their product. At least in North America, the tobacco campaign is pretty much over, but that on global warming still has legs. One of the major reasons for the ongoing controversy on climate change is that the worst expected consequences will not become unmistakably obvious for decades to come.

Global warming is a matter of greenhouse gases (GHGs) in the atmosphere, the most important of which are carbon dioxide, methane, and water vapour (black carbon particulate matter is a separate risk factor). Delayed impacts are a part of the intrinsic nature of global warming, because some GHGs emitted into the atmosphere, primarily carbon dioxide from fossil-fuel use (coal, oil, and natural gas), do not disappear or quickly fall back to earth again; some considerable part of them stay there for long periods of time, creating the "greenhouse effect," which will be explained later. This fact gives rise to the important distinction between greenhouse-gas *emissions*, on the one hand, and *concentrations*, on the other. Emissions are the gases which intermittently come out of our chimneys and automobile exhausts each day; concentrations are the more permanent and stable levels of GHGs in the atmosphere, which, under present conditions, tend to rise slowly year over year.

Some of these delayed impacts of global warming are, quite literally, centuries in the making. In calculating the threat of rising greenhouse-gas concentrations in the atmosphere, for example, scientists often refer to the point when those concentrations will have doubled since the beginnings of the Industrial Revolution around year 1750. Moreover, unlike marching armies and millions sick and dying from disease, we cannot definitively "see" this particular threat itself. Both greenhouse gases and their effects on climate are invisible, and thus the link between GHGs and climate is not obvious. Even if we were to seek to become aware of their presence, we would possibly discount the threat, because, we would learn, they make up only a few percentage points in the atmosphere we breathe.

Like the gases themselves, the danger they represent is not readily apparent to us. Scientists now tell us that, first, we are at risk of crossing over a threshold of warming that, again, we cannot observe directly. Second, once having crossed, we may be unable to go back. Third, when we finally can see the worst things appear right before our eyes, we may learn both that we can no longer avoid them and that they may continue to worsen *for centuries to come*. We can list these possibilities for the global impacts of climate change in more technical language as follows:

1   The world may have only about another decade, until around
    2030, to begin taking seriously the threat of crossing a thresh-
    old of danger by starting to drastically reduce levels of green-
    house-gas emissions.
2   If we do cross the threshold of danger, it may not be clearly
    apparent to us that we have done so; it is possible that this will
    occur as early as around 2050.
3   If we do actually cross this threshold, there may be *nothing*
    we can do to reverse the course of events, except perhaps to
    attempt to engineer the global climate (which will itself carry
    significant risks).
4   The worst outcomes, such as severe and persistent droughts,
    coastal flooding, damage to our forests, and numerous other
    impacts, may not begin to occur until quite a bit later in this
    century; major social disruptions may start at this point.
5   Sometime in the second half of this century the world may
    become locked into a future in which the serious impacts will
    steadily worsen for centuries to come.

Special note should be taken of the use of the word "may" in these five propositions. These are probabilities, not certainties, and it can be difficult to put any precise numerical figure on how likely it is that these threats will occur. But scientists have for some time now asserted that their main conclusions about climate change are held with "very high confidence."

Nations of the world have been trying to get themselves onto a consistent pathway leading to reductions in their greenhouse-gas emissions for the last thirty years. With some exceptions, notably in the European Union, they have largely failed to do so up until now. Politicians have postponed taking these actions in part by setting deadlines for achieving them that are far in the future. Citizens in many cases have been reluctant to incur the economic costs (such as carbon pricing) for now, since there does not seem to them to be any urgency in doing so. The bottom line is that humanity may fail utterly to address this risk, despite the fact that the seriousness of the likely future consequences for not doing so is already clearly described – and that there may be little time left in which to change course.

In this book I seek to assess briefly the situation that faces us now, both in global and in specifically Canadian terms. There are just two main aspects that need examination. Risk is a combination of probability and consequences. Thus, the first question has to do with what we think we know about the risk, asking: How likely is it that very bad things could happen, as a result of climate change, at some time in the future? And the second is, what kinds of bad things could occur, and just how bad might they be? We can try to answer such questions only by consulting the consensus reports of the climate-science community around the world, since they are, quite simply, the *only* available resource for useful and reliable information on this matter.

Then we need to review the attempts by nations to arrive at their own kind of consensus about what political actions are required in order to respond to the conclusions and guidance in those scientific reports. Those attempts have primarily sought to use the mechanism of an international treaty, the United Nations Framework Convention on Climate Change (UNFCCC), which came into force in March 1994. Under this umbrella, some or all countries make promises to each other about what decreases they will engineer with respect to their greenhouse-gas emissions. What we need to know above all is whether and to what extent nations have delivered on their promises, both generally around the world and particularly in Canada.

When those two aspects of the climate-change situation have been clarified, we will be in a position to realize how well we have done to date, and how much more remains to be accomplished in order to fulfill the UNFCCC goal of avoiding "dangerous anthropogenic interference with the climate system."

These objectives dictate the succession of short chapters to follow here. In chapter 1, I introduce the concept of "climate," which must be understood clearly in order for anyone to see what we are concerned about with respect to climate change. In chapter 2, I suggest that knowing something about the two most recent periods of climate history, known as the Pleistocene and the Holocene, during which modern humans evolved, helps us to grasp the issues confronting us now. In chapter 3, I summarize the development of modern climate science, which extends back into the nineteenth century, and in particular science's use of "general circulation models" in its attempt to predict the future course of climate change. In chapter 4, I make the case for putting our trust in the climate story that science tells. In chapter 5, I look at the contemporary story of international negotiations among nations seeking agreement on objectives and the responsibilities for fulfilling them; in particular, I focus on Canada's record of participation in those negotiations and its record of achievement in living up to the promises it has made to the international community.

In chapter 6, I take a detailed look at the gap between the greenhouse-gas emissions targets derived from climate science, on the one hand, and, on the other, the results to date of the international treaty negotiations that seek to fulfill those targets. In chapter 7, I analyze some of the causes for the failures to date in the international treaty negotiations on climate change and review the most important options for the future: decarbonization, geoengineering, and carbon management. In chapter 8, I examine where Canada stands in doing its share in reducing greenhouse-gas emissions. I look at the targets it has set for 2030 and 2050, encompassing mitigation (actions that may inhibit further warming of the climate) as well as impacts and adaptation (changes made to institutions so that they can adjust to climate warming).

I restate here the guiding theme that I intend to follow through the chapters to follow. It is this: both a full public understanding of the scope of the risk that climate change represents, and the willingness of the public to support the measures taken by their governments to control that risk, depend on one simple proposition – namely, that climate change is unlike every other risk with which they are familiar. In a nutshell, what is different about climate change risk is that the descendants of people who are alive today might not experience the full measure of the catastrophes which will descend on the nations of the world until it is far too late for them to be able to do anything to prevent those tragedies.

The chief difficulty many citizens have with this proposition is that they must put their trust in a scientific account of the risk that is exceedingly complex and, for most of us, almost impossible to fully grasp on account of its complexity. During 2021 and 2022, the Intergovernmental Panel on Climate Change (IPCC) has been issuing the newest in its series of comprehensive reports, known as the Sixth Assessment Report (AR6). It consists of many thousands of pages of dense and demanding text and references from the published academic literature. Very few outside the community of climate scientists will ever read or study it in its entirety. Policy-oriented officials who serve political leaders will consider themselves fortunate if they can just get through the "Summary for Policymakers" that accompanies the scientific reports.

Most of us do recognize clearly how much we depend on the insights and technologies bestowed on us by the modern sciences to sustain our lifestyles. Yet there is a natural human inclination against making sacrifices now for benefits which will only be realized at some point in the future. We tend to apply what is called a "discount rate" to future benefits, thinking that they are worth far less at present than they will be eventually, so why should we pay for them now? We think that "the future will take care of itself," that earlier generations have overcome serious challenges such as pandemics and wars, and therefore future generations can be trusted to do the same when it comes to climate change. After all, we imagine, people will have better technologies at hand in the future to deal with problems than we do now. If we think about climate change risk at all, we tend to suppose that, *if it really is as serious as scientists think it is*, then our children and grandchildren will buckle down and deal with it.

Climate change is, then, a great test for those of us who are living in the present, for actions we do take – or alternatively fail to take – in the coming three decades will likely determine whether or not certain types of serious catastrophes may befall our descendants. Our scientists tell us today that such catastrophes are "very likely" (viewed with "high confidence") to occur unless we begin from this year forward to devise and support the policies that may enable future generations to avoid them. Therefore, we have a clear choice before us: we can believe what scientists say, even if we cannot fully understand the reasoning behind their messages, or we can just take our chances – on behalf of the coming generations – that they will be proved wrong. This is a bet. How good a bet is it?

Posing this question means, in effect, that each Canadian has two choices. *The first choice* is to say, "I will support national policies that require my country to reach net-zero greenhouse-gas emissions no later than the year 2050." If this is your position, then you will expect those national policies to adopt the most economically efficient measures available for controlling GHG emissions in this country. As explained later, these measures will be based on policies such as carbon pricing, which can be constructed in such a way as to represent a modest incremental financial burden on you and your family. Supporting this position means that you will take one side of the bet: namely, that what the great majority of Canadian climate scientists are telling you about global warming is reliable and trustworthy.

The second choice is to say, "I'm not sure that human-caused climate change is actually happening or that it will have very bad consequences for my children and grandchildren." If this is your position, you will not support national policies that are designed to halt or mitigate global warming and any expected impacts. It means that you do not believe that harmful global warming is likely to happen, or if does, it will be the result of natural processes, not human actions, and thus it is something beyond our control. Supporting this position means that you will take the other side of the bet: namely, that what the great majority of Canadian climate scientists are telling you about global warming is unreliable, and quite likely either untrue or just mistaken.

This is the bet now in front of you. *You must take one side or the other*: your neighbours and fellow citizens will ask you to place your bet, because Canada's national and provincial governments are already putting the questions behind that bet on the table. So, which side of the bet will you take? I ask the reader to pose this question for himself or herself after reading this book.

1

What Is Climate?

The most important [climate] processes have been different over these different time scales – changes in the sun have been offset by changes in greenhouse gas consumption by rock weathering over billions of years, continental drift has altered patterns of atmospheric and oceanic circulation and of greenhouse gas production and removal over hundreds of millions of years, and features of Earth's orbit have affected the distribution of sunlight, ice, and greenhouse gases over hundreds of thousands of years.

Richard Alley, *The Two-Mile Time Machine*

Our earth's climate system is a prodigious mechanism of energy transfer. This energy transfer drives other processes that, taken together, produce what is now known as the planet's climate. One of its essential features is the sheer length and duration of its past phases, extending in total over four billion years of the earth's history, phases which have occurred in terms of hundreds or tens of millions of years, and sometimes of millennia and centuries. This feature of our earth – the long past history of its climate, and its major phases – is a relatively recent discovery of modern science: Until about two hundred years ago, it was entirely unknown to us.

"Climate" is something we cannot see or feel; it is a *concept*, an idea, constructed both out of now-abundant evidence and complex theories and reasoning. What we see and feel each day is *weather*, which fluctuates daily and seasonally. *Climate*, on the other hand, is the record of much-longer-term phases that can differ very markedly from each other. The scientific account of climate tells us, for example, that some fifty-five million years ago palm trees grew in the Arctic. We know this because scientists found pollen buried deep in the ocean floor there and estimated the time during which the trees flourished by using isotopic dating.

Looked at from this angle, "climate" is no different from any of the rest of the scientific concepts developed in the modern age. To list just a few of them: in chemistry, atoms and molecules; in physics, radioactivity, the photoelectric effect, and quantum mechanics; in biology, genetics and the evolution of species. All these concepts are indispensable for crafting the understanding of the natural world that gives rise to the production of goods and services sustaining our lives and keeping us safe. We don't need to understand them in order to enjoy their fruits. However, we all must grasp what climate is, because, in the absence of this knowledge, we will not be able to comprehend the meaning of what is currently called "climate change," or why it is important that we should deal with it adequately in terms of our public policies – or what might happen if we do not.

The energy transfer mentioned above arises from two quite different sources: first, internally, from heat rising from deep within the earth's core and mantle, and second, externally, as a result of solar radiation emanating from our sun. During the 4.57 billion years of our earth's history, the first source has gradually diminished; at the same time, heat from the sun has slowly increased by about 30 per cent, reflecting its growing luminosity, a standard phase of the known life cycle of what is classified as a yellow dwarf star. A third but intermittent cause, also an external one, has been strikes from large comets and asteroids, primarily occurring in the earliest period of the earth's history. On average, large asteroids have struck the earth once every hundred million years, the last being the most famous, sixty-six million years ago, which created the Chicxulub crater beneath the Yucatán Peninsula in Mexico and was a factor in one of the great extinction events of biological life. As for comets, there is a plausible theory that large bodies of this type, which are composed of rock, dust, and ice, struck the earth during the first two billion years of its history and were the source of the water in our oceans.

When considering climate as a mechanism of energy transfer, we need to take brief look at the two primary and enduring sources of that transfer: stored energy and solar energy.

The huge amount of heat still stored in the earth's interior originated during what is called the development of the proto-earth, which was an accretionary process of a gravity-fed accumulation of dust and gas lasting for five billion years. In addition, much of the earth's residual heat at its core is a result of a Mars-size planet, named Theia by scientists, which collided with the earth about 4.5 billion years ago, whereupon the two bodies combined. (Mars is one-sixth the size of Earth.) This generated a massive wave of heat in the earth's crust, melting it to a depth of thousands of kilometres and ejecting enough matter to form our moon.

Even today the temperature of the earth's inner and outer cores is about five thousand degrees Celsius. This heat radiates continuously from the interior to the exterior of the planet, but its chief impact on the surface results from volcanic activity in the earth's upper mantle, driven by super-hot magma (molten rock), which seeps up from deep below and which appears as lava on the surface. Two other relatively minor types of stored energy are frictional heat in subterranean processes and the decay of radioactive elements.

The second main source of energy transfer is solar radiation. Sunlight strikes the earth in the form of electromagnetic radiation, consisting of infrared, ultraviolet, and visible light. Our sun is a main-sequence star, which formed out of a cloud of gas some 4.6 billion years ago. It will remain in this phase for about another 3.5 billion years before it slowly begins to heat up and expand, becoming a red giant, then cooling to end up as a white dwarf star. About 6.5 billion years from now, the red giant phase of the star in our solar system will have roasted our earth's surface into bare rock, dotted with glowing lakes of liquid metals, and will have evaporated all the water in our oceans.

The history of volcanic activity on our earth is not a story of peaceful change, but rather one of extraordinarily violent activity, driven by the stores of residual heat in the earth's mantle. The most visible manifestations of this violence are, of course, volcanic eruptions and earthquakes, which are now understood as a function of plate tectonics. The earth's crust is composed of a collection of vast, intersecting platforms, or plates, on which the continents and the oceans sit, and they grind against each other, pulling apart and pushing against their boundaries. This discovery of the earth's composition is a splendid twentieth-century scientific achievement based on the use of seismic waves, whose shape and speed as they propagate through the planet provide clues to what lies below our feet. Scientists now believe that plate tectonics play a critical role in both climate modulation and the generation of the earth's magnetic field.

Scientists have found evidence of past supermassive volcanic events – for example, at the Siberian Traps, which are called "large igneous provinces," extending back in time some three billion years.

These events, as well as less massive volcanic eruptions, have had a direct effect on the composition of the atmosphere and the climate. The events that are most relevant to human life occurred about sixty-six million years ago, a combination of two distinct factors: first a large asteroid colliding with the earth, ejecting an immense amount of debris into the atmosphere; and second an igneous province in India known as the Deccan Traps. The result was a dramatic cooling of the climate, which resulted in the loss of 75 per cent of the extant species on the planet and marking the end of the Cretaceous Period. All the non-avian dinosaurs were among the victims; their disappearance allowed mammals to flourish, including eventually our own species.

The most direct impact of the planet's composition on biological life is its effect on the atmosphere, which is held in place by gravity and is stratified into layers, from the densest near the surface to the thinnest, the exosphere, the boundary between the atmosphere and outer space. In the earth's earliest history, the atmosphere was first composed mostly of hydrogen gases, such as ammonia and methane. The second phase, beginning about four billion years ago and occurring during the heavy bombardment of the earth by huge asteroids, was made up of nitrogen and carbon dioxide. This gave rise to the carbon cycle, and this phase also included what is known as the Great Oxygenation Event, starting some 2.45 billion years ago. Now comprising 21 per cent of the atmosphere, oxygen fuels the metabolism of animals and plants. One or two episodes, during which the earth was almost totally covered in ice, snow, and slush, occurred some 750 to 550 million years ago (MYA), the second of which lasted 100 million years – but which, happily, was followed by the "Cambrian Explosion," a huge expansion of animal and plant life forms.

Earth is a planet where physical processes in its subsurface formations remain powerfully active and have a huge impact on its surface. An example is the movement of the tectonic plates upon which continents reside. Some of these processes result in a massive, never-ending recycling of carbon between surface and subsurface. In the earth's early history, some four billion years ago, when huge amounts of carbon dioxide were being released by volcanic eruptions, the surface temperature of the earth may have been as high as 60°C (140°F). This was our planet's introduction to what is called the "greenhouse effect," which is one of the three great "blankets" that protect biological life here.

The first blanket is the planet's magnetic field, the magnetosphere; in its absence (as is the case on Mars), the stream of high-velocity charged particles emanating from the sun, known as the solar wind, would strip away our atmosphere. The second is the ozone layer in the upper atmosphere, which protects plants and animals from excessive amounts of ultraviolet radiation. And the third is the greenhouse effect, which refers to the fact that the earth's surface is warmed by absorbing radiation from the sun, mostly in the visible wavelengths, some of which is reflected off the surface, especially by glaciers and sea ice, and reradiated back into space in the form of long-wave infrared or heat radiation. However, fortunately for us, some of this reflected heat energy is trapped by the atmosphere. It is like living in a greenhouse. But unlike the clearly visible greenhouse in our backyard, whose warmth we can feel upon entering, even in wintertime, the climatic greenhouse effect is not apparent to us; before the studies of some nineteenth-century scientists (Joseph Fourier, John Tyndall, and Svante Arrhenius), we were entirely unaware of it.

This third blanket depends on the atmospheric concentration of the various gases there, notably water vapour and ozone produced by natural processes, which absorb infrared radiation and warm it. Water vapour is particularly important, owing to the fact that as the atmosphere warms it can hold more of it, creating a positive feedback loop. Then there is another group of greenhouse gases (GHGs), which includes carbon dioxide, methane, nitrous oxide, and some fluorinated compounds. All except the last of these can be created by natural processes, but also independently by human activities, and thus are human-caused (anthropogenic). This other group of GHGs is what is referred to in the policies and agreements that seek to control the human production of greenhouse gases.

The fact that the earth's average surface temperature at present is about 14°C (57°F) is due to this heat-trapping effect, without which the surface temperature would be a full 32°C colder (–18°C or –0.4°F). The average temperature at the earth's surface was much warmer at specific times in the distant past than it is now, reaching 8°C (14.4°F) hotter than today some fifty-five million years ago and steadily declining since then to 6°C (11°F) below present levels some twenty thousand years ago, before it rose again to the current level.

This chapter opened by characterizing the climate system as a powerful energy-transfer mechanism on our planet. "Climate" itself is defined as the mean and variability of certain indicators, especially temperature and precipitation, over periods of time ranging from thirty years to hundreds of millions of years. Scientists estimate the climatic condition in the earth's distant past from evidence such as tree rings, cores drilled into ocean sediments (containing residues of animal and plant materials), fossils preserved in rock formations, and ice cores drilled into glaciers, using isotopic dating to figure out the time sequence. The clearest lessons drawn from this evidence is that climate is dynamic, not static, and that climate has varied greatly throughout the earth's history. Only one aspect of these ever-changing conditions is seasonal and hemispheric variability, which are functions of shifts in the earth's orbit around the sun and the tilt of its rotation axis (known as eccentricity and obliquity). An excellent introduction to the complexity of the earth's climate and the many forms of heat transfer that influence it will be found in an online course "Earth in the Future" (Balower and Bice 2022). For example, there is the well-known "thermohaline circulation," called the "great conveyor belt," which moves massive quantities of heat through deep ocean currents from the tropics to the northern hemisphere.

The earth's climate, therefore, is a changing sequence of conditions *played out over very long periods of time*. Human lives are often acted out in terms of seconds, minutes, and hours, as well as days and years, up to centuries and millennia, whereas climate can be measured in cycles of a hundred thousand or a hundred million years. As a result, thinking about climate requires us to reset drastically our very notion of meaningful time. Climate is characterized by predominant patterns in key indicators such as global average temperature, the gaseous composition of the atmosphere, the effects of water and rock (erosion), volcanic activity, solar radiation, and the interactions of all these factors. The most fundamental point about climate is that it refers to long periods having various sets of these indicators which fluctuate widely within certain ranges. Changes in climate have occurred many times in the earth's history, transitioning from one phase to another, often slowly but sometimes more quickly as a result of massive asteroid strikes or huge volcanic eruptions.

In the distant past there were long periods when the average temperature on Earth was both very much colder than it is now (14°C) and sometimes considerably warmer. For example, during the Cryogenian Era (also known as "snowball earth"), dating from about seven hundred million years ago (MYA) and lasting for more than a hundred million years, the global mean surface temperature was as low as −50°C and glaciers blanketed the northern hemisphere as far south as the equator. But there have also been long periods when the earth was much warmer than it is now, such as toward the end of the Neoproterozoic, around six hundred MYA; the Cretaceous Hot Greenhouse (ninety-four MYA); and the Paleocene-Eocene Thermal Maximum (fifty-six MYA), when

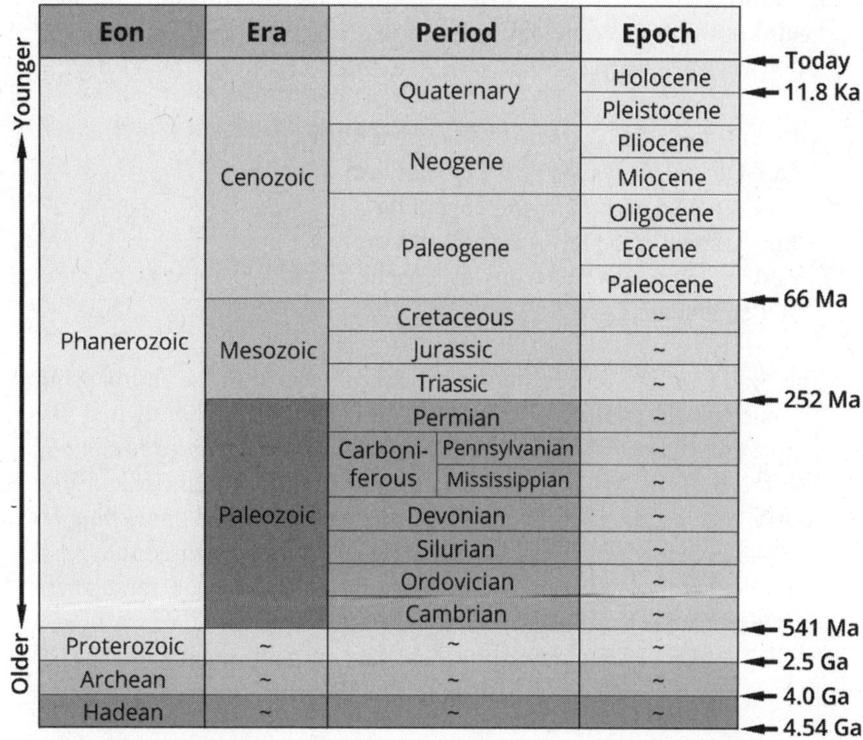

Eon	Era	Period		Epoch	
Phanerozoic	Cenozoic	Quaternary		Holocene	← Today / ← 11.8 Ka
				Pleistocene	
		Neogene		Pliocene	
				Miocene	
		Paleogene		Oligocene	
				Eocene	
				Paleocene	← 66 Ma
	Mesozoic	Cretaceous		~	
		Jurassic		~	
		Triassic		~	← 252 Ma
	Paleozoic	Permian		~	
		Carboni-ferous	Pennsylvanian	~	
			Mississippian	~	
		Devonian		~	
		Silurian		~	
		Ordovician		~	
		Cambrian		~	← 541 Ma
Proterozoic	~	~		~	← 2.5 Ga
Archean	~	~		~	← 4.0 Ga
Hadean	~	~		~	← 4.54 Ga

(Left axis: ← Younger / Older →)

Figure 1.1 | Geological epochs.
Ka = thousands of years ago
Ma = millions of years ago
Ga = billions of years ago

average surface temperatures were +30°C, crocodiles swam in the Arctic Ocean, and sea levels were considerably higher than they are now. Like the earlier hot periods, the more recent one was caused by an extreme greenhouse effect involving huge releases of CO_2 and methane due to massive volcanic eruptions and the melting of frozen methane hydrates in the oceans.

To sum up: life on Earth is thought to be 3.5 billion years old, but complex life forms date from just the past one-half billion years. Fossil and other evidence show how changing climatic and environmental conditions – including shorter-term episodes – have determined the prospects for various plant and animal species. Scientists specializing in the new fields known as paleobiology and geochronology have documented the following five events, known as "mass extinctions," in Earth's history:

End Ordovician, 444 MYA, 86 per cent of species lost;
Late Devonian, 375 MYA, 75 per cent lost;
End Permian, 251 MYA, 96 per cent lost;
End Triassic, 200 MYA, 80 per cent lost;
End Cretaceous (Cretaceous–Paleogene boundary), 66 MYA, 76 per cent lost.

The "End Permian," occurring at the boundary between the Permian and Triassic periods, is the one known colloquially as "the great dying." The simple fact is that particular species *are and must be adapted to climatic conditions during specific periods in the earth's history.* This includes, of course, our own species, modern humans, whose rise and flourishing on this planet extends back in time no more than three hundred thousand years during the Quaternary, an insignificant fraction of our planet's 4.57-billion-year existence.

The sequence of geological epochs is shown in figure 1.1.

The Pleistocene and Holocene Epochs

The Quaternary is the most recent of the three periods of what is known as the Cenozoic Era, "the age of new life," and it extends from about 2.6 million years ago (MYA) to the present day. It thus covers the entire span of time in which the ancestors of modern humans, and after them we ourselves (*Homo sapiens*), evolved in Africa, and is therefore the specific era of the earth's ever-changing climate history that is most meaningful to us today. The Quaternary is composed of two epochs, the first of which is called the Pleistocene, lasting from 2.6 MYA to about 11,800 years ago; the second, the Holocene, beginning 11,800 years ago, is the climate epoch most directly relevant to the period when large groupings of settled human societies created all civilizations.

Thus the important timeline for the relation between human history and climate history extends from the late Pleistocene through the Holocene. Since the period of the middle Quaternary, about one million years ago, there have been a regular series of events lasting one hundred thousand years each and referred to as the glacial-interglacial cycle. In figure 2.1, take note not just of the dominant pattern (the eight major cycles) but also of the constant smaller variations across the whole record.

Periods of extensive glaciation have occurred throughout the entire Pleistocene era, but about a million years ago there developed a repeating glacial-interglacial pattern on a hundred-thousand-year cycle. This pattern results from complex variations in our planet's tilt on its axis and its orbit around the sun, known as the Milankovitch Cycle. These variations affect the distribution of solar radiation striking the planet's surface. The actual mechanism driving the glacial-interglacial cycle is the variation in solar insolation – that is, the amount of sunlight striking the earth's surface – in the northern hemisphere. During the

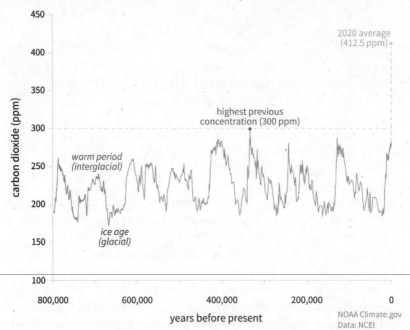

CARBON DIOXIDE OVER 800,000 YEARS

Figure 2.1 | The glacial-interglacial cycle and CO2 levels.
Global atmospheric carbon dioxide concentrations (CO2) in parts per million (ppm) for the past eight hundred thousand years. The peaks and valleys track ice ages (low CO2) and warmer interglacials (higher CO2). During these cycles, CO2 was never higher than 300 ppm. On the geologic time scale, the increase (vertical dashed line) looks virtually instantaneous.

most recent Glacial Maximum, the ice reached as far south as 40° latitude (about where Portland, Seattle, Chicago, and New York City are now located) and was as much as four kilometres thick. These events consisted of roughly ninety thousand years of cold weather (an "Ice Age") in the northern hemisphere, dominated by immense glaciers, and an interglacial period averaging ten thousand warmer years between each glacial period.

During the glacial periods, the atmosphere over much of the planet was cold, dry, dusty, and windy. The regular transitions from glacial to the warmer interglacial occurred during the earth's axial precession, which is the changing orientation of its rotational axis over a period of about twenty-five thousand years; this type of phenomenon can be observed in a top that is set spinning on a table. More sunlight in the northern hemisphere began to melt the glaciers, and then the warming of the oceans released stored carbon dioxide, which amplified the warming in the atmosphere. The complexity of the climate impact of the axial precession is shown in the temperature record, which indicates that there were warming surges during the glacial periods that were, however, insufficient to trigger the beginning of a full interglacial.

The climatic record of the last eight hundred thousand years is deduced from the ice cores drilled into Antarctica and Greenland, yielding round vertical columns of ice up to three kilometres in length (Davies 2020). The accumulated ice had been built up over millennia in annual increments of snowfall; the pressure created during the ice sheet's buildup forms tiny bubbles in the ice, made of gases such as carbon dioxide, oxygen, and methane, which when measured provide evidence for the concentrations of those gases in the atmosphere at the time when a particular annual increment was laid down in the column. The long ice-core records also, among other things, provide additional definitive evidence for the close association between temperature and carbon-dioxide levels.

This point is important in the present context, because it illustrates the variability in the underlying geophysical processes that cause the climate to change. As we shall see a bit later on, at present scientists believe that the addition of huge volumes of carbon dioxide to the atmosphere by human activities is the main driver of climate change now. But the glacial cycle shows that this is not always the case. Since the earth's climate is so variable, what is important is to understand the specific mechanism at work at any one time.

One of the major conclusions that can be drawn from the discipline of paleoclimatology is that the earth's climate has been in fact quite "wildly variable," to use the words of Richard Alley, a glaciologist who spent decades studying the past history of Greenland ice as part of a team drilling two miles into its glacier and extracting ice cores. In chapter 8 of his important book *The Two-Mile Time Machine*, written for the general public, Alley discusses one of the great paradoxes of our earth's climate history:

> Climate can be rather stable if nothing is causing it to change, but when the climate is "pushed" or forced to change, it often jumps suddenly to very different conditions, rather than changing gradually.

You might think of the climate as a drunk: When left alone, it sits; when forced to move, it staggers ... Over long times, Earth's feedbacks act to oppose the forcing, so large causes produce small effects. Over shorter times, Earth's feedbacks amplify the forcing, so small causes have large effects.

(Alley's full explanation of this paradox is highly recommended.) The reader should keep this idea in mind for some of the discussions to follow in this book, particularly the material on how glaciologists use the ice-core drillings from Greenland and Antarctica to date the stages in past climates. They found evidence that shorter-term changes can be both abrupt and very large. For example, temperatures ascertained in measurements in ice cores from Greenland and dated between fifteen thousand and twelve thousand years ago reveal episodes when they first rose 15°F over a period perhaps as short as a decade, then plunged again, and finally jumped once more by the same amounts. We will want to recall this point later when the issue of the abruptness of recent human-caused forcings of the climate is reviewed (see Brovkin et al. 2021, for the most recent study of abrupt climate change in the past thirty thousand years). In the climate-change context, a "forcing" is some factor that causes the climate to change. Positive forcing, such as solar radiation, warm the climate; negative forcing, resulting for example from the gases released during volcanic eruptions, cool the climate. As discussed later, there are both natural and human-caused types of forcings (Colose et al. 2020).

The second important scientific investigation in recent times is into the isotopic dating of fossil remains, through which scientists can track the evolution and dispersal of anatomically modern humans throughout the world. The last glacial retreat marked the beginning of the Holocene, which means that today we are well beyond the midpoint of the current interglacial. And this is the time when the modern human population exploded on the planet. It is estimated that, seventy thousand years ago, during the last glacial era, there were no more than ten thousand modern humans anywhere in the world. Population estimates for the onset of the Holocene range from one to ten million, although there are large uncertainties, and the number is very likely to be at the low end of that range. Our distant ancestors, species of archaic humans – first *Homo erectus* and then *Homo heidelbergensis* – began dispersing out of Africa as much as two million years ago. *Homo heidelbergensis*, which flourished about five hundred thousand years ago, was the probable progenitor of our close relatives, the Denisovans and Neanderthals.

Anatomically modern humans arose in Africa as much as three hundred thousand years ago and began leaving some seventy thousand years ago, first heading East to Asia and Oceania, then to Europe about forty thousand years ago. Our own species (*Homo sapiens*), along with our Neanderthal and Denisovan cousins, endured and then began to flourish throughout the last three in a series of glacial-interglacial cycles, each lasting about a hundred thousand years. During the last Ice Age, modern humans occupied parts of northern Eurasia as the continental glaciers waxed and waned; average temperatures were about 6.1°C colder than today.

The tenuous relationship between earlier humans and changing climatic conditions is illustrated well by what happened during and after the period known as the Late Glacial Maximum (LGM), occurring between twenty-seven thousand to nineteen thousand years ago, which was marked by a severe cooling of the climate and the expansion of the continental ice sheets. Anatomically modern humans were already well settled in Europe at the onset of that period, but this population suffered a serious decline as a result of the climatic change and was forced to retreat to the southernmost areas of Europe. There was some significant climate instability just before the LGM, and this may well have been a factor in the extinction of our cousins, the Neanderthals.

Following the LGM there was a repeated shifting between shorter-term warming and cooling phases, as the climate system was in the process of transitioning from the last glacial to the latest interglacial. In this context, "shorter-term" means periods of one to a few thousand years. The transition from the glacial to the interglacial may be likened to attempting to start a combustion engine that has been sitting idle for a very long time. On the initial tries, the engine turns over but fails to catch. Around 14,500 years ago, during the rapid onset of one of the severe cooling episodes, the existing human population in Europe was basically wiped out, to be replaced later, when temperatures rose again, by a distinctively different group. The evidence for this process relies on mitochondrial DNA retrieved from fossil remains.

The East Antarctic ice-core results present a picture of the temperature and CO_2 record across eight glacial-interglacial cycles. For dating dealing specifically with the Holocene (covering only the most recent 11,800 years) there is a trend line of rising global temperatures following the Late Glacial Maximum, when around 20,000 years ago the temperature was 6°C (11°F) colder than it is now. But there were also significant intermittent episodes of cooling, especially in the period called the Younger Dryas (10,000 to 8,500 years ago); two notable "cold events" during this period are linked to large pulses of cold fresh water pouring into the North Atlantic from the melting Laurentide ice sheet, disrupting the oceanic heat transport, known as thermohaline circulation, from the equator to the poles. Ice cores from the Greenland ice sheet, which is some two kilometres thick, provide the most precise data for the Holocene, showing that there has been a remarkable degree of climate stability, beginning about eight thousand years ago and lasting until relatively recently. Figure 2.2 shows the temperature climbing out of the end of the last glaciation in two distinct phases.

Domestication of plants and animals in agriculture and herding is thought to have begun some thirteen thousand years ago, just before the onset of the Holocene, and one estimate puts the total human population at two million around 10,000 BCE. Following the Younger Dryas, shorter and less severe cooling cycles alternated with warming ones. From 5000 to 3000 BCE, the Holocene Maximum experienced temperatures 1 to 2 degrees Celsius (1.8 to 3.6 degrees Fahrenheit) above the current level, ancient civilizations flourished in Egypt and elsewhere, and the human population rose to forty-five million.

Like their hominid ancestors, earlier humans were scattered bands of hunter-gatherers. A very different lifestyle, beginning about thirteen thousand years ago and based on the domestication of plants and animals, led to the growth of settled societies in the regions of what is now the Middle East starting some eight thousand years ago. The early permanent settlements gradually expanded into more complex societies, and by 3500 BCE we see the emergence of the first human civilizations in Mesopotamia (modern Iraq) and Egypt. The period from eight thousand to three thousand years ago is known, in climate terms, as the "Holocene Climatic Optimum." The first system of writing – cuneiform – also dates from this time, although there are forms of record-keeping on clay tablets for counting and recording goods and property that are somewhat older.

Modern humans survived the last Ice Age, but in small numbers and mostly in the Southern Hemisphere, which was not as cold as the Northern but still considerably cooler than at present. The happy coincidence in time between the arrival of a particular geological climatic period (the Holocene) – which was welcoming to warm-blooded upright mammals – on the one hand, and, on the other, the earlier evolution of a primate species (*Homo sapiens*) primed to exploit and even enhance the life-sustaining resources found at hand in its environment was truly a fateful throw of nature's dice. The geological history of this specific

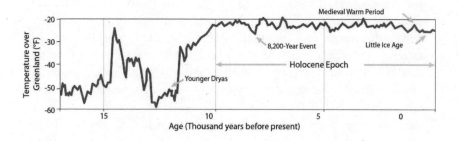

Figure 2.2 | Holocene climate.

planet, violently and repeatedly refashioning its crust and atmosphere across eons of time, and the complete evolutionary history of biological life on its surface, billions of years in the making, joined forces at precisely the right time to set the table for modern humans, allowing us to show how much we could do with the opportunity that nature had bestowed upon us.

The warm Holocene undoubtedly raised the productivity of agriculture and the size of domestic herds. A dramatic increase in food supply resulted in steady population increases; a mid-range estimate for the population in 3000 BCE is fourteen million, compared with perhaps just two million at 10,000 BCE. In another millennium it had doubled again, and by year 0 there were about 170 million. During what is known colloquially as the "Little Ice Age," a long cooling period lasting from about 1300 to 1850, global average temperatures decreased about 1°C (1.8°F) from the level reached in the Medieval Warm Period. During the early stages in this cooler period human population growth ceased or declined somewhat, as a result of such events as the Great Famine and the Black Death in Europe in the early fourteenth century, but the overall trend line for the human population for the last two millennia has been relentlessly upward, reaching a milestone of one billion people for the first time around 1800, then leading to exponential growth in the twentieth century. By the end of 2020, the total stood at 7.8 billion individuals.

The period when the human population began rapid and sustained further growth after 1800 also marked the beginnings of the Industrial Revolution. This was not a coincidence, because industrialism brought technologies that revolutionized human life. The production of one such innovation in particular, namely the Haber-Bosch process for the synthesis of nitrogen from air, achieved in the early decades of the twentieth century and creating the ammonia used in artificial fertilizer, stands out. There are credible estimates that the effect of this single innovation on the food supply is responsible for half of the population increase since the beginning of the large-scale production of ammonia. Then there were the public-health innovations, especially sanitary measures and chemicals for disinfection, medicines, and vaccines, prior to which half of all infants died before the age of five and pregnancy and childbirth presented the greatest risks in a woman's lifetime. Thanks primarily to a vast increase in the food supply and the control of infectious diseases, the doubling time of the human population after 1800 dropped from 127 years to about 50, with the eight-billion level expected to be reached by 2023.

The overall significance of this epoch in climate history can be summed up by saying that *human civilizations have been well-adapted since their beginnings to the climate conditions of the Holocene.* Although factors such as average temperature, precipitation patterns, sea levels, and others continued to fluctuate somewhat, they did so within a relatively narrow range. At present, we have begun to move outside this range and to do so at an accelerating rate, with no end in sight. As will be seen in the following chapters, climate scientists have advanced the hypothesis that warming to 2°C above pre-industrial levels (year 1750) and beyond may very well set in motion a series of climate-change impacts, such as dramatic sea-level rise, that over time could be catastrophic for us. The human species undoubtedly would survive such a transition, but much of its socio-economic and industrial as well as its cultural frameworks may then lie in tatters.

There is a strong association between the gradual emergence of the first human civilizations and an extended period (in human terms, at least) of a favourable climate. *But that relatively stable climate has now ended.* Changes in carbon dioxide and temperature levels track each other, and CO_2 levels are now at a concentration in the atmosphere of 417 parts per million. Temperatures are very likely to continue rising in the present century, even if there are no further increases in CO_2 concentrations (although there will certainly be such increases).

In 2000, the chemist Paul J. Crutzen, who had won a Nobel Prize for his contribution on the ozone-depletion issue, popularized the term "Anthropocene," referring to it as the period – dating from the onset of the Industrial Revolution – during which our species has become so dominant on the planet as to be responsible for a transition to a new planetary epoch. In this new epoch, the major threats to other life forms, occasioned by human-caused habitat destruction and other factors, involve loss of biodiversity, sharp declines in the population of wild land animals and amphibians, destruction of rainforests and boreal forests, and oceanic acidification. Recent scientific estimates about the magnitude of the accumulated human impacts on the biosphere, expressed in terms of biomass, are: (1) of all mammals now on Earth, 60 per cent are livestock, 36 per cent are humans, and 4 per cent are wild; (2) chickens and other poultry are 70 per cent of all birds, the remaining 30 per cent are wild; (3) since the beginning of human civilization, 83 per cent of wild land mammals and 80 per cent of marine mammals have disappeared. A single species is well on its way to crowding out most of the other wild animal life on the planet. If evolutionary fitness for an animal species is defined – from one perspective, at least – as an ability to dominate its environment, to become what is called a "top predator," then we, *Homo sapiens*, have become the worthy successor of the non-avian dinosaurs. Perhaps this is why many find *Tyrannosaurus rex* so fascinating.

The sum total of all human impacts on the environment has been called our species' "ecological footprint." Our total demands placed on the store of natural capital (natural resources) can be assessed with respect to the criterion of *sustainability*. How likely is it that the current level of demands on resources by the population that exists now, and by further human population increases, can be satisfied from both the planet's regenerative biocapacity and its stock of depleting stores? And for how long into the future? (To be sure, the intensity of average per-capita demands varies widely across the spectrum of richer and poorer nations.) A consolidated image of our ecological footprint is presented in the idea that at present "1.7 earths" are necessary in order to satisfy the total human demands placed on our planet's environmental resources. This means that our present level of demands exceeds the earth's capacity to satisfy them sustainably – that is, indefinitely into the future – and that we are quickly drawing down the accumulated natural capital of the earth – its bioproductivity and stock of non-renewable resources.

This consolidated image also leads to the question of whether all these accumulating human impacts may result in what is known as an ecological collapse, involving a sharp and perhaps sudden reduction in existing biological productivity across the planet as a whole, constraining its carrying capacity for all extant species, including our own. Major events of this type are known from the geological past, especially the mass extinctions previously listed, which were caused by events such as violent and prolonged volcanic eruptions, large asteroid impacts, and sudden climate change.

Recently other scientists have been exploring the concept of "planetary boundaries," a set of nine discrete parameters designed to measure the resilience of the earth's chief biogeophysical systems that sustain human life under the present conditions. Their analysis starts with the following observation (Steffen et al. 2015): "The relatively stable, 11,700-year-long Holocene epoch is the only state of the ES [Earth System] that we know for certain can support contemporary human societies." They then ask whether the Holocene earth system can persist in the face of current human pressures against it, as assessed by measurements in nine dimensions: atmospheric aerosol loadings, altered biogeochemical cycles, biosphere integrity, climate change, freshwater use, land-system change, novel entities, ocean acidification, and stratospheric ozone depletion. They regard two of the nine (biosphere integrity and climate change) as "core," or critically important, processes. They find that, in a total of four of these nine (biogeochemical cycles, biosphere integrity, climate change, and land-system change) – a set which includes both of the core dimensions – human activity may already be pushing us beyond the boundary zone, the point where it becomes uncertain whether the earth system that now sustains our species can persist.

We know that our earth has undergone many extensive geological transformations since its origin. Even if we accept the proposition that humans have now embarked on a pathway to the future that may undermine the established foundations of their present way of life, possibly drastically so, this means nothing with respect to the entirety of the earth itself. The planet's atmospheric and geological processes will adjust, as they always have done, and will transition into some new state of equilibrium. The larger-scale processes known to have occurred in the Late Quaternary, that is, the repetitive glacial-interglacial one-hundred-thousand-year cycles, either will persist long into the future, until there is a transition to a different state, or they will be disrupted relatively soon and transition more suddenly to the next state, whatever proves to be the case. In either event, the planet will carry on, except that there may be a new mass extinction of many extant species, a type of event that has taken place a number of times in the distant past. We know that the remnants of life have picked themselves up thereafter and carried on in new ways. But this time the history of the entire human effort to erect complex civilizations is at stake.

3

Predictions of Climate Science

> It is unequivocal that human influence has warmed the atmosphere,
> oceans and land ... The scale of recent changes across the climate
> system as a whole and the present state of many aspects of the climate
> system are unprecedented over many centuries to many thousands
> of years.
>
> IPCC – AR6, *Summary for Policymakers* (2021)

The first volume of the Sixth Assessment Report (AR6) from the Intergovernmental Panel on Climate Change (IPCC), *Climate Change 2021:
The Physical Science Basis* – four thousand pages long – appeared in
August 2021. Its forty-two-page *Summary for Policymakers* (SPM), the
document that almost everyone in the public-policy sphere will peruse,
opens with this statement: "It is unequivocal that human influence has
warmed the atmosphere, oceans and land." That statement has never
appeared before in any of the previous five assessment reports and has
been made in the context of the judgments of all the world's leading
climate scientists that there are increasingly dire consequences on the
near horizon.

The story about climate so far is that relatively recent scientific techniques and discoveries have given us an increasingly sharp picture of
climate changes over the past eight hundred thousand years, to which
has been added the less detailed – but still robust – accounts of earlier
phases. The scientific reconstructions of past climatic epochs, including quite distant ones, rely necessarily on appropriate evidence. In an
earlier chapter, I noted that geologists had developed ways of demonstrating that episodes of massive volcanic eruptions occurred ten times
in the past three billion years. The Cryogenian Episode, an epoch during
which scientists believe that northern-hemisphere glaciers extended as

far south as the equator, has also been mentioned. In this case, some evidence consists in the discovery of "sedimentary structures that could have been created only by glacial activity" in the deposits left behind in equatorial regions as glaciers retreated again.

EVIDENCE OF PAST CLIMATE AND
PREDICTIONS OF FUTURE CLIMATE

With the increased understanding of the more recent record of past climatic history, two things in particular prompted scientists to ask a different set of questions. First, there was the close association over time between average temperatures and the atmospheric levels of CO_2 during the later Pleistocene, as shown in the four-hundred-thousand-year record from the Vostok ice core from East Antarctica (Alley et al. n.d.). Second, there was the evidence that levels of CO_2 in the atmosphere were steadily rising during the twentieth century, from 296 parts per million (ppm) in 1900 – as measured in ice cores – to 370 ppm in 2000, as measured directly at the Mauna Loa Observatory in Hawaii. Third, the rate of increase was accelerating, tripling from an average of 0.74 ppm/year throughout the twentieth century to an average 2.25 ppm/year in the period from 2000 to 2020. Taken together, these factors and some others convinced many climate scientists that they should focus on using newer techniques in attempting to *predict the course of climate change into the near future.* The rest of us need to grasp at least the outlines of why and how they do so. This story begins with the discovery of the greenhouse effect in the nineteenth century.

Studies by Joseph Fourier in the 1820s advanced the hypothesis that the earth's atmosphere warms our planet. These studies were followed by the path-breaking work in the 1860s of John Tyndall, who first investigated the capacity of various gases in the atmosphere, including nitrogen, water vapour, carbon dioxide, and ozone, to absorb heat. Tyndall also showed that the process of absorption of radiation from the sun changes its electromagnetic character from short-wave to long-wave (infrared) radiation, so that some of the resulting heat is trapped by the atmosphere and is not reradiated back into space – although the now-familiar term "greenhouse effect" was not coined until 1901.

Toward the end of the nineteenth century, the Swedish scientist Svante Arrhenius focused for the first time on the actual quantities of the two key drivers of this process, water vapour and CO_2, in the atmosphere, and pointed to the combustion of fossil fuels as one of the main elements in increasing the levels of carbon dioxide. Furthermore, Arrhenius did the first quantitative studies of the potential impacts of the growing carbon-dioxide levels in the atmosphere, predicting that a doubling of these levels would result in an average temperature increase of 5°C, which is remarkably close to modern estimates. His famous ground-breaking paper, which refers to carbonic acid (carbon dioxide dissolved in water), entitled "The Influence of Carbonic Acid in the Air upon the Temperature of the Ground," was published originally in German in 1896 and in English translation that same year.

In this section of the chapter, only a few key developments in the understanding of the greenhouse effect during the latter part of the twentieth century are summarized. In 1957, Roger Revelle and Hans E. Suess published a journal article entitled "Carbon Dioxide Exchanges between Atmosphere and Ocean and the Question of an Increase of Atmospheric CO_2 during the Past Decades," which included a passage that became famous:

> Thus human beings are now carrying out a large-scale geophysical experiment of a kind that could not have happened in the past nor be reproduced in the future. Within a few centuries we are returning to the atmosphere and oceans the concentrated organic carbon stored in sedimentary rocks over hundreds of millions of years.

Shortly thereafter, in 1958, Charles David Keeling, who like Revelle and Suess was a member of the Scripps Institution of Oceanography in La Jolla, California, established the Mauna Loa Observatory in Hawaii, at an elevation of three kilometres, and began taking direct measurements of atmospheric concentrations of carbon dioxide. The result is now known as the "Keeling Curve," a steadily rising line with annual seasonal variations.

A mere seven years later, Keeling's findings were included for the first time in a United States government report and were the basis there of both a warning for the future and policy prescriptions on pollution control. This was the 1965 President's Science Advisory Committee Report on Atmospheric Carbon Dioxide, entitled "Restoring the Quality of our Environment," which contained the following sentence: "By the year 2000 the increase in atmospheric CO_2 will be close to 25%. This may be sufficient to produce measurable and perhaps marked changes in climate and will almost certainly cause significant changes in the temperature and other properties of the stratosphere." It was also the first government document to predict that there would likely be some dramatic impacts from unrestrained levels of growth in atmospheric carbon dioxide, including the melting of ice caps and rising sea levels.

The report's "Conclusions and Findings" included a restatement of the key passage appearing in the 1957 Revelle and Suess journal article:

> Through his worldwide industrial civilization, Man is unwittingly conducting a vast geophysical experiment. Within a few generations he is burning the fossil fuels that slowly accumulated in the earth over the past 500 million years. The CO_2 produced by this combustion is being injected into the atmosphere; about half of it remains there. The estimated recoverable reserves of fossil fuels are sufficient to produce nearly a 200% increase in the carbon dioxide content of the atmosphere.

This analysis is worth reproducing here, because it is important for citizens today to understand that *warnings about unconstrained rises in atmospheric CO_2, as well as its likely adverse impacts, have been made for well over half a century*. Since then, starting four decades ago and carrying on to the present day, large groups of scientists have sought to reiterate this message and reinforce it with new research. The first result was the 1979 report from the US National Research Council, *Carbon Dioxide and Climate: A Scientific Assessment*; this was followed by much larger efforts undertaken by the Intergovernmental Panel on Climate Change (IPCC), founded in 1988 by the United Nations Environment Program and the World Meteorological Organization, which has produced forty-seven substantial reports, beginning in 1990 and continuing down to the present.

COUPLED GENERAL CIRCULATION MODELS (CGCMS)

A number of different types of models are used to describe the climate system. One of these produces energy balance models (EBMs), which are limited to calculating surface temperature and are based on three factors: solar input, albedo (reflection of energy radiated back into space), and the chemical composition of the atmosphere. Another major and more complex type produces "general circulation models (GCMs)," which seek to include all of the physical processes occurring in or on the oceans, the atmosphere, the land surfaces, and the cryosphere (ice-covered areas of the planet). One group of such models, known as "coupled GCMs," attempt to unify all atmospheric and ocean general-circulation elements.

CGCMs are the most comprehensive models, integrating all the physical processes that influence climate, and for that reason they are difficult to construct.

Coupled general circulation models are imaginative reconstructions of the earth's climate system made up of four dimensions, consisting of three spatial dimensions plus time. These form a grid, akin to sets of boxes piled above and below each other, one set for the earth's surface, one for the oceans, and one for the atmosphere. The atmospheric grid may have as many as twenty vertical layers and the oceanic, thirty layers. Enormous amounts of data generated by the whole set of boxes are input into the model, which is why running such models requires the use of the largest supercomputers available. The latest versions of these models incorporate the following types of data: interactive vegetation; dust, sea spray, and carbon aerosols; upper atmosphere; atmospheric chemistry; atmospheric/land surface, oceans, and sea ice; sulphate aerosols; biogeochemical cycles; carbon cycle; marine ecosystems; and ice sheets. In more familiar terminology, the inputs include measurements of such factors as water vapour, solar radiation, wind, clouds, ocean circulation, albedo (or reflectivity off ice and snow), heat, atmospheric gases, and others.

The great complexity of the models is a function of the fact that what is happening in each of the three key spatial components (land surface, atmosphere, and oceans) continuously interacts with the others, as do some of the separate factors, which means that all the positive and negative feedback loops among them must be described and measured. The CGCMs use equations drawn from the principles of physics, notably thermodynamics and fluid dynamics – such as the law of conservation of energy – to specify how these interactions occur. These interacting elements are then translated into lines of code using a programming language, usually FORTRAN, which works with the familiar IF/THEN/DO routines, totalling over one million lines of code and taking up close to twenty thousand pages of text for a full global model. The basic unit of analysis is the "grid cell," which might have a spatial resolution of 100 km squared in terms of latitude and longitude.

The results are simulations or re-enactments of the complex natural processes which, scientists believe, give rise to the climatic events which we actually experience in real life, such as rain, heat, cloud cover, wind, and storms, as well as other processes that we cannot observe directly, such as the flow of carbon among atmosphere, land, and oceans. The outputs from running such models can give a day-by-day representation of how the earth's climate changes over long periods of time. The UK's

Hadley Centre and other agencies use the same models for short-term weather forecasting. (For an excellent Canadian journalistic account of this scientific work see Fairbank 2021.)

In sum, CGCMs are extraordinarily complex constructions made up of interacting large-scale processes (such as the hydrological cycle and the carbon cycle), huge data sets of many different kinds (for all the separate factors), and analytical methods drawn from physics and chemistry. In order to validate the results that they generate, scientists look to see whether their models provide a generally acceptable level of agreement with the known and measured climate and weather conditions of the past 150 years. They seek to fine-tune their models by varying certain parameters and rerunning them again and again. When they are satisfied that the model's predictions of past events are as close to what actually occurred as they can achieve, they run the models forward in time to make predictions about what is likely to happen in the future. The results are probabilities, that is, estimates of how likely it is that specific events will happen, and their objective is to achieve high confidence in those predictions.

RESULTS: WHAT IS THE MAIN FACTOR IN CLIMATE CHANGE TODAY?

Most climate models are run from about 1850 to the present. As noted above, the models are built using physical processes, such as the carbon cycle, and then are filled out with certain data inputs, the most important of which are the amounts of energy from the sun, but also forest fires and volcanic eruptions and concentrations of greenhouse gases (these are collectively known as positive and negative "climate forcings"). They are designed to produce many specific outputs, notably temperature, humidity, snowfall, rainfall, wind speed, and the extent of glaciers and sea ice. Once the model is up and running, scientists use "hindcasts" (the opposite of forecasts) – testing the models against past temperatures – to calibrate its accuracy. In other words, since we have some actual measurements of climatic conditions going back one hundred years or more, scientists can observe the extent to which the model predictions correlate with actual measurements and then work on fine-tuning the models until the correlations are as accurate as possible.

Scientists can also put a wide variety of hypothetical data into their models, such as a potential doubling of greenhouse-gas concentrations at some time in the future, and see what model outputs they get. Models

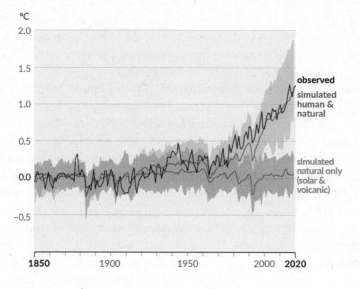

Figure 3.1 | Models of change in global surface temperature using both natural and human forces.

are also run for past climates, such as the Holocene or the Paleocene-Eocene Thermal Maximum, the warming period of 55 MYA. Finally, it should be noted that there are about a dozen fully-fledged coupled models in use at research centres around the world, as well as a specific program, a coupled model intercomparison project (CMIP), to test them against each other.

Figure 3.1 is, quite simply, the one decisive graphical representation that everyone concerned about climate change needs to understand and appreciate. Scientists created it by first running their models using only natural climate forcings, especially solar radiation, and then comparing the outputs with recent measurements, such as average temperature and the atmospheric concentration of greenhouse gases. They then estimated the amount of forcing attributable to humans, especially in the past two hundred years, notably greenhouse-gas emissions from industry and land-use changes. The difference between the two, that is, between "models using only natural forces," on the one hand, and "models using both natural and human forces," on the other, is shown on the right-hand edge of the graphic in figure 3.1. It first appeared in the IPCC Fourth Assessment

Report (2007) and has now been superseded by a more complete diagram, Figure SPM-1, Panel b): "Changes in global surface temperature over the past 170 years," shown in the first section of the *Summary for Policymakers* as part of the Sixth Assessment Report, *Climate Change 2021: The Physical Science Basis* (see also EPA 2022 and NAP 2020).

The most significant general finding from climate science is that, over the course of the twentieth century and continuing down to the present, anthropogenic (human-caused) activities are the main reason that global temperatures appear to be rising relentlessly (Semeniuk 2021a). There are a number of such actions, such as land-use practices, but by far the most important is the release of increasing amounts of greenhouse gases, especially carbon dioxide and methane, as a result of human activity, and the burning of fossil fuels stands out as a decisive factor. In this regard scientists emphasize the concepts of climate forcing and climate sensitivity, that is, the extent to which the earth's global average temperature changes in response to increases in the emissions of greenhouse gases. Beginning in the late 1980s, groups of climate scientists have advised governments and their citizens to institute policies that would rein in the emissions of these gases, primarily by moving away from generating energy by fossil-fuel use and by mandating the use of alternative sources of energy, such as solar, wind, and nuclear power.

Like the end-states that emerge from the operations of all very large and complex systems, both natural and human-constructed, the future trajectory of the earth's climate cannot be easily diverted. In this respect, the climate system is rather like human societies themselves, which for the most part respond to new information and changed environmental conditions slowly at best. As we have seen, scientists want to know how the earth's climate system will respond over the longer term to the induced energy imbalance resulting from the human-caused loading of greenhouse gases in the atmosphere. They know that other factors will influence this response, in a set of both positive and negative feedback loops: water vapour, clouds, and sea ice, for example.

The process which interests scientists most is what the climate system's response would be (such as future temperature changes) to an expected doubling of the concentration of carbon dioxide in the atmosphere since the onset of the Industrial Revolution, with its greatly enlarged use of fossil fuels, in the late eighteenth century. Predicting when the climate will respond to this specific input raises the problem known as *thermal inertia*. Even if new inputs, representing human-caused emissions of these gases, were somehow to be halted at once and completely, considerable

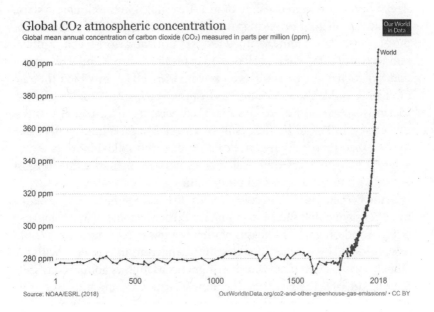

Figure 3.2 | Carbon dioxide in the atmosphere since year 0.

time would elapse before the climate system eventually reached a new equilibrium level in response to this change. Thermal inertia is related to what is called the "atmospheric residence time" of various gases, which is the amount of time during which a gas continues to react to solar radiation, trapping energy and causing the atmosphere to heat up as a result. In simple terms, this means that, were we to decide at some point to try to stop the earth's temperature from continuing to rise by reducing inputs of anthropogenic greenhouse-gas (GHG) emissions, the positive impact of our decision, a halt in rising temperature, would not be registered in the atmosphere until some decades thereafter.

In this context, climate scientists began referring to specific "thresholds" in the global-warming scenarios, for two primary reasons: (1) thermal inertia, as described; and (2) the risk that, after a certain amount of warming had been induced by anthropogenic GHG emissions, some natural positive feedback loops might come into play, the most consequential of which would be the release of huge quantities of methane – a potent greenhouse gas – that for now remain sequestered in

Arctic permafrost and ocean hydrates. Thresholds in the climate system, such as the melting of permafrost and glaciers, represent possible tipping points, that is, some attained levels of critical factors (such as global temperature) that, when exceeded, may result in abrupt and irreversible *additional* changes, possibly even a "runaway" effect, in which the rate of change suddenly accelerates and cannot be brought under control. (Scientists regard the extreme surface temperature of the planet Venus – 462°C or 864°F – to be the result of a runaway greenhouse effect due to the exceptionally high concentration of carbon dioxide in its atmosphere.) As with every other calculation in a risk scenario, this forecast comes with uncertainties and probabilities.

Some people who live in cold climates may respond to these scenarios by either saying that such a warming would be welcome news or, alternatively, wondering why increases of merely a few percentage points could be considered by scientists to constitute dangerous interference with the climate system. Rebecca Lindsey comments as follows about today's situation: "In fact, the last time the atmospheric CO_2 amounts were this high was more than 3 million years ago, when temperature was 2°–3°C (3.6°–5.4°F) higher than during the pre-industrial era, and sea level was 15–25 meters (50–80 feet) higher than today." The point at which the strong and persistent "uptick" began to occur is the arrival of the Industrial Revolution around 1750. Since global GHG emissions are still rising as of 2020, this rise will inevitably be translated into an increase in global average temperatures. A 1.3°C (2.33°F) rise over pre-industrial levels has already occurred (Berkeley Earth 2021), and with current trends there is a risk that the climate system may become locked into a 1.5°C (+2.7°F) increase quite soon, sometime between 2020 and 2030. Does it matter that a 1.5°C rise would exceed the upper boundary of the temperature variation that is estimated to have occurred during the entire Holocene, the period during which all human civilization developed?

Unless actions are initiated soon to begin reducing anthropogenic greenhouse-gas emissions to eventually stabilize the concentrations of these gases in the atmosphere (that is, preventing them from continuing to rise), a global average temperature increase of 2°C (3.6°F) above pre-industrial levels may occur well before the end of the twenty-first century. Still, this can appear to be a small increase, so does it really matter? And if so, why?

Just how serious might a 2°C global temperature increase be? Might a 2°C global warming be the level at which humanity unavoidably would be set on a course for a catastrophic future? A scientific paper (Steffen et al. 2018) begins as follows:

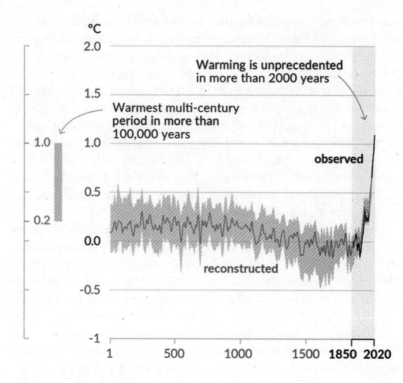

Figure 3.3 | Global warming.

We explore the risk that self-reinforcing feedbacks could push the Earth System toward a planetary threshold that, if crossed, could prevent stabilization of the climate at intermediate temperature rises and cause continued warming on a "Hothouse Earth" pathway *even as human emissions are reduced* [italics added]. Crossing the threshold would lead to a much higher global average temperature than any interglacial in the past 1.2 million years and to sea levels significantly higher than at any time in the Holocene. We examine the evidence that such a threshold might exist and where it might be ... Where such a threshold might be is uncertain, but it could be only decades ahead of a temperature rise of [around] 2.0°C above preindustrial.

According to these scientists, passing the +2°C (+3.6° F) temperature threshold might set in motion what they call "tipping cascades," which are positive biogeophysical feedback loops (permafrost thawing, loss of sea ice, release of frozen methane from oceans, and others) that accentuate the trends in rising temperatures already occurring. Potential catastrophic effects following +2°C include a sea-level rise of as much as six metres, severe reductions in food output, and extensive dieback of both boreal and tropical forests. But the even more serious possibility is that, once at an increase of 2°C, the climate system may become locked into the "Hothouse Earth" pathway, causing additional temperature increases, a trend that could well be irreversible, the effects from which would persist for millennia thereafter. This open-access article by Steffen et al. has beautiful and instructive graphics; readers are strongly encouraged to access them and study them carefully.

The greatest risk is that humanity may turn out to be unable to mount effective countermeasures to avoid the dangerous "Hothouse Earth" pathway because the current *rates of change*, for both atmospheric CO_2 concentrations and temperature, are so high: Steffen et al. (2018) write that "these current rates of human-driven changes far exceed the rates of change driven by geophysical or biosphere forces that have altered the Earth System trajectory in the past." These human-driven changes even exceed the events which brought about the Paleocene-Eocene Thermal Maximum, occurring some fifty-five million years ago, when global temperatures were 8°C (14.4°F) higher than they now are. Faced with a high rate of change leading toward certain end points, any efforts we make to counteract the trend must be initiated sooner rather than later, scientists argue, or else we face rapidly diminishing opportunities to alter the trajectory of future events. As another group of scientists (Aengenheyster et al.) put it in a 2018 paper, we may be approaching the "point of no return" in climate change, the point at which we no longer have the option of avoiding future rising temperatures and catastrophic outcomes.

4

Trusting Climate Science

According to an analysis of the promises made by nations in the latest updating of the Paris Agreement, at the Glasgow meetings in November 2021 (CAT 2021), the world is heading for something like a 3°C warming over pre-industrial levels. Paleoclimatologists tell us that the last time the world was 3°C warmer than it is now, about three million years ago, sea levels were seventeen metres higher than at present. Yes, the climate on planet Earth has always changed. Perhaps there are even times when it would not have changed fast enough for us, for example, if we humans had stood at the onset of the snowball-earth episodes and were facing the next hundred million years or so of frigid weather. But that was seven hundred million years ago and fortunately there were no mammals about. Neither the evolution of species, nor the environmental niche in which we have flourished so dramatically, had yet prepared the preconditions for our arrival and success as a species. Chi Xu et al. (2020) write: "All species have an environmental niche, and despite technological advances, humans are unlikely to be an exception. Here, we demonstrate that for millennia, human populations have resided in the same narrow part of the climatic envelope available on the globe, characterized by a major mode around 11°C to 15°C mean average temperature."

This is the Holocene, our climate cradle. Chi Xu et al. go on to say that "this temperature niche is projected to shift more over the coming 50 years than it has moved since 6000 BP." Six thousand years before the present encompasses the entire period during which human civilizations came into being and during which the human population expanded from an estimated eleven million to a projected 9.7 billion in 2050.

Could environmental conditions really shift enough in the coming half-century to abruptly and roughly force us out of the supportive climate niche we have enjoyed for the past six thousand years? Why should we believe that such a change is likely or even very likely to begin to happen then or sometime later in the present century? Even if these events were to occur, could things really become so bad as to jeopardize the entire way of life we have come to enjoy? Finally, even supposing we were to entertain the idea that all of this might take place at some point in the future, why should we Canadians believe that we could avoid such a disastrous future by adopting only one very specific strategy, right now, this year – namely, to begin reducing our GHG emissions and not stopping until we have reached our target of net-zero emissions by 2050? Even if we accept that we should make changes to reduce our emissions somewhat, why should we believe that we need to take any such drastic steps by 2050? Why can't we wait a bit longer, until we can be more certain that either we don't have to bother or that we really have no choice other than to embark on this strategy?

In order to provide well-reasoned and evidence-based answers to questions such as these, many hundreds of scientists from around the world, drawn from a wide variety of academic disciplines and based in many different countries, have collaborated for over five decades on extremely detailed overall assessments in climate science. Published papers on these subjects in peer-reviewed journals easily number in the thousands, even in the tens of thousands. The analytical methods they employ in this area are drawn from the shared common stock of knowledge inherited from their predecessors over the past few centuries. The methods used by climate scientists are in every respect similar or identical to those used in every other contemporary scientific venture of discovery in physics and chemistry, notably thermodynamics and atmospheric chemistry.

The questions posed just above and others like them have a single answer: We should accept what we have to do because, over the past fifty years, the most accomplished natural scientists, both in Canada and around the world, hundreds and indeed thousands of them, have said so. Analyses have shown that there is a robust 97 per cent consensus in published climate-research studies that human-caused GHG emissions are the primary cause of recent global warming and of expected further warming in the coming years. In order to facilitate the efforts of readers who wish to examine carefully the evidence on this point for themselves, in the references section for this chapter I have listed the open-access URLs for six relevant published articles. Readers may also inspect the references given in those articles, which include a few papers challenging the specific claims advanced in some of them. To the best of my knowledge, however, there are no published articles in reputable, peer-reviewed academic journals that contradict the claim made about the 97 per cent consensus position.

The essence of the entire scientific argument is contained in a single graphic, which has been shown earlier (figure 3.1). The case that there is a demonstrated difference between natural climate forcings on their own, and forcings where human effects are added to the mix, means that we humans have assumed responsibility for the temperature rise since 1950 – and that further such rises are confidently expected if present levels of GHG emissions continue to increase.

Why should the rest of us, who are not climate scientists, believe this contention? So-called "skeptical" attitudes range from the view that human actions cannot possibly be a decisive influence on the planet's climate to a questioning of scientists' motives. The awkward difficulty resulting from these views is that, if the methods employed by climate scientists are erroneous or impure, then so in equal measure are the findings of all their colleagues in related fields, since the latter use exactly the same scientific methods. Another awkward truth is that this common, shared method of the investigation of nature underpins all the technologies and medical devices that these same doubters utilize and appreciate in daily life. The fact that these and other devices usually work as intended, confirming daily the truth of the scientific methods that make them possible, is something all of us experience every day of our lives.

A scenario about the future that is probabilistic in nature, as all risk scenarios are, tells us that something harmful might occur later on unless steps are taken right away to head it off. It is not unreasonable, when faced with such a prediction, to ask whether one might wait for more certainty before acting. Whether or not this would be a prudent thing to do depends on the nature of the risk, however. Applying the "wait-and-see" approach in the case of the climate system may be dangerous, for in delaying too long actions that are needed to reduce the risk, we might arrive at a point when the harmful events cannot be avoided, no matter what we do then or thereafter.

This kind of bold and alarming prediction should give us pause. And then we might ask ourselves: could the entire large group of scientists, living in many different countries around the world, be just plain wrong about climate change? We might think that this is by no means an unreasonable question. After all, the history of modern science demonstrates that leading scientists of their day have occasionally been wrong about important points in their various disciplines. In physics, as late as toward the end of the nineteenth century, there was a widespread adherence to the theory of the "luminiferous aether," supposedly an invisible medium through which light was propagated. It doesn't exist. In chemistry, there was the phlogiston theory, used for about a century to explain combustion until being rejected in the late-eighteenth century. And, until about the same time, naturalists believed that biological life-forms were fixed and did not evolve. Finally, throughout the eighteenth century, competing schools of thought in geology battled against each other.

One reason for the existence of these long-running controversies is that, before 1900, communications among scientists living in different nations were infrequent and slow and confined primarily to Western Europe. However, since the end of the nineteenth century, the population of working scientists has increased enormously and has also expanded around the globe. Since the creation of the World Wide Web, large-scale scientific collaborations on projects and papers across great distances occur daily in great numbers. The daily communications, frequent meetings, and joint publishing ventures among them have also been greatly enlarged. These and other factors make it much less likely that major interpretive errors will take root, persist, and remain unchallenged in any scientific discipline today, as compared with the past.

Also, the multidisciplinary character of the climate-science field is one of the attributes that protects it from developing major interpretive error. For example, thermodynamics is one of the oldest core areas of modern science; it overlaps the fields of both physics and chemistry, and it is also an indispensable element in many modern technologies, including engines. Thermodynamic equations are used by climate scientists in their coupled general-circulation models, and it would be easy for the thermodynamics specialists who work in subfields other than climate studies to tell if the uses of those equations in these models are either inadequate or erroneous. There is no plausible suggestion that they are.

To be sure, there is no doubt that in certain respects science remains incomplete down to the present day: There are lively debates about the nature of physical reality in its smallest dimensions; the standard model of particle physics remains incomplete; relativity and quantum mechanics are not unified; and all physicists would love to know what dark energy and dark matter are. Much more remains to be understood in biochemistry (such as protein folding) and genetics (such as DNA repair) as well. There is reason to speculate that studies in the natural sciences, like other intellectual and artistic endeavours, will never be finished. *And yet incompleteness, unsolved puzzles, and unresolved disagreements over specific points of interpretation are not the same thing as major interpretive error.*

Further, the climate-science community, like all scientific groupings, continually refines and improves the theories and methods it employs and develops new sources of relevant data. So, at any moment, one can assume that there are some as-yet-undiscovered shortcomings in their collective work that will be overcome by climate scientists themselves at some point in the near future. But is it possible or even likely that the current consensus among scientists seeking to explain climate change might turn out to be wrong in its entirety? Less provocatively, we might appropriately ask: Even if one were to accept fully the contention that the earth has been warming somewhat since the middle of the twentieth century, and that this warming is accelerating, could there be some simple alternative explanations for these observed changes? For example, could they have resulted from purely natural processes, such as increases in solar radiation or something else? An answer is given in the major climate-science consensus documents, one of which is the *Climate Science Special Report*, issued in 2017 by a group of leading US scientists and available in its entirety on the Internet:

> Over the last century, there are no convincing alternative explanations supported by the extent of the observational evidence. Solar output changes and internal natural variability can only contribute marginally to the observed changes in climate over the last century, and there is no convincing evidence for natural cycles in the observational record that could explain the observed changes in climate. (*Very high confidence*).

The evidence for this statement is shown in figure 4.1.

Of course, the authors of this report could be wrong, but if so, where is the contrary evidence? Or worse: Have they been deliberately perpetrating an elaborate hoax on all the rest of us – and on their colleagues in all other fields of science? This is a bit like the assertion that the 2020 US presidential election was "rigged" or "stolen." None of the proponents of these claims ever explains *how* such elaborate undertakings in fakery could actually have been carried out. A question relevant in the present context is: *How* could any climate science hoax be perpetrated, since all scientific studies are in the public domain and are subject to appraisal by

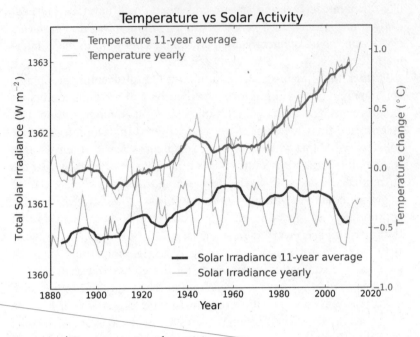

Figure 4.1 | Temperature vs solar activity.

the scientific community as a whole? The modern scientific consensus on anthropogenic climate change has its origins in the famous 1957 journal article by Roger Revelle and Hans Suess, and has been regularly reaffirmed for over sixty years. If, sometime in the coming decades, this same scientific community investigating the climate comes upon new data and develops new theories that call into question either or both the concept of human climate forcing and the perceived need to drastically reduce anthropogenic greenhouse-gas emissions, we will know about these developments *only* because they will have been evaluated and published in the academic literature.

To contend that such contrary research findings, should they ever be made, could or would somehow then be suppressed, or that the current scientific consensus on climate change amounts to a gigantic hoax, is simply an irresponsible and groundless charge. It is true that, very occasionally, an individual scientific paper that has been peer-reviewed and then published in a reputable journal contains misrepresented or even invented data and is subsequently withdrawn, and that sometimes this even amounts to academic fraud. But it is *impossible* to imagine that this could occur on the scale of the thousands of papers on climate science that have been published since 1957. It is likewise impossible to imagine that all their authors have simply invented the whole problem, and that at some point it will just go away of its own accord. To accept either of those propositions is to call into question not only the integrity of the entire process of modern scientific investigation since its sixteenth-century origins, but also the evidence that exists before our very eyes, the evidence that our science-based technologies do actually work.

In short, it is frankly preposterous for anyone to contend that, in all the vast domain of the modern natural sciences, *every single subfield yields reliable and useful results with the sole exception of climate science.* This simply cannot be true.

We depend on our scientists to explain to the rest of us how and why the world around us operates as it does. There are many comments in the preceding sections which call attention to one crucial aspect of the modern sciences, namely, that they must wrestle with the fact that the greater part of the reality of nature remains hidden – and deeply hidden – from our ordinary senses. The ways in which nature's many different operations actually produce the experiences of what we see and feel in the world around us are screened from view. The instruments devised to unmask these unseen realms began with simple telescopes and microscopes and advanced ultimately to the incredibly complex particle colliders and coupled general-circulation models of today.

Common sense asks: How can it be that the solidity of the material objects we handle every day is an illusion, because the particles which compose matter are mostly just empty space? How can it be that an invisible electromagnetic force, known to science as simply "the strong force," holds together the constituent particles that make up atoms? How can it be that the world as it appears to our eyes is only a small part of what is happening in the universe, because the full electromagnetic spectrum contains many other dimensions – for example, ultraviolet and infrared radiation and X-rays – that we cannot see without the aid of specialized equipment? All this and much more may be decidedly odd, when considered from the standpoint of common sense, but it is not possible to doubt, as we sit waiting for our MRI and CT scans, that what scientists tell us about these phenomena is true.

The scientific study of our earth's climate is another mystery of this same type. We cannot "see" the climate; what we see and feel and hear is the daily weather. Scientists must *construct* past climate history from the many inferences they draw out of the huge troves of evidence that are stored in the geological history of the earth: rocks, ocean sediments, tree rings, long cores drilled from the ice sheets, fossilized plant and animal remains up to six hundred million years old, and other data. They can tell us, for example, that palm trees once grew in an ice-free Arctic region some fifty-three million years ago, when the climate there was like Florida is today, because they have found palm pollen in sediments on the ocean seabed just five hundred kilometres from the North Pole. They can tell us what the atmosphere and the oceans were like hundreds of millions of years ago, because isotopes of oxygen and carbon are preserved in the shells of tiny creatures called foraminifera and diatoms. But we cannot look around ourselves in our neighbourhoods and see the climate history of the earth. That story is told in the planet's geological and atmospheric history as it has been reconstructed by generations of scientists. On the basis of that history, they have also made some educated guesses as to what the near future might hold for us.

Climate scientists carry on with their work as some governments dither about what response they should make to it. Sooner or later, governments around the world, especially those in the nations that are the largest emitters of greenhouse gases (China, the United States, and a few others), and their citizens (assuming they have a voice in the matter), will have to decide either to accept the scenarios and predictions summarized above or to ignore them – as they have the legitimate authority, and the legal right, to do. Climate scientists have provided a sense of the probabilities of the harms that await us, as well as the level of confidence they have in those numbers. To be sure, it is possible that they have misinterpreted or exaggerated both the likelihood and the consequences inherent in the risks of climate change. The key questions for the rest of us are: How certain are we that they have exaggerated or misinterpreted the risk? If we are just doubtful about what to believe, we might then ask ourselves: How long can we wait before making up our minds about what these scientists are saying?

For some, climate-change skepticism means refusing to believe what is asserted in the consensus view of scientists and choosing to accept what is read and heard through other sources, although very few among the rest of us have the knowledge and skills needed to independently evaluate the validity of alternative viewpoints. This skepticism does appear to be eroding, and as of now even a strong majority of citizens in the United States report to pollsters that they are convinced about the reality of global warming. However, this amounts to only the first baby step toward a determination to support policies and actions robust enough to bring about an end to rising GHG emissions. A conviction that resolute action of this kind is not only desirable but necessary means that citizens must pay the full economic and social costs required to make it happen – and many of us, even those in countries whose elected national governments support the appropriate public-policy measures, still appear to be a long way from that next step.

It is also important to understand that all risk scenarios come with uncertainties, no matter how elaborate the basis of the relevant scientific knowledge, and climate-change risk is no exception to this rule. The earth's climate is a highly complex geophysical system, and it is to be expected that there are some uncertainties related to it. This by no means justifies the view that we should wait for more certainty before taking the necessary actions to reduce global GHG emissions. On the contrary, a well-reasoned argument has been made by Wagner and Zeckhauser (2018) to the effect that climate-change uncertainties make it more imperative, not less, to implement those steps now.

Some experts say that, even as of 2020, there was very little time left during which we could make a meaningful difference in future outcomes by curbing emissions growth. To be sure, there is some remote possibility that, before any important deadlines or thresholds have passed, the scientific consensus may change, and they will tell us that we need not go to the trouble of reining in our GHG emissions at all. How likely is it that waiting for this possible change in scientific opinion is a wise thing to do? Our sitting and waiting for this eventuality is nothing less than an ongoing wager on our future. It is a bet on how likely it is that any dramatic change in the current scientific consensus on climate forcing will occur well *before* we embark irrevocably on a Hothouse Earth pathway. Because if and when that happens, it then becomes likely that nothing at all we attempt to do in curbing GHG emissions will make any difference to future outcomes.

Therefore, our doing nothing now, or not enough to make a difference, or too late to do so, can be framed as a wager not only on our own future but on that of our children and grandchildren. The youngest cohort of people alive today will very likely begin to experience, during their lifetimes, some of the more serious impacts of climate change. Perhaps most of those alive today will have passed away before the worst of the predicted adversities become apparent, but before that many of them may come to realize that nothing they can do will avoid the eventual appearance of climate catastrophes. Their children and grandchildren will be the ones required to reflect on how wise or unwise it was for their ancestors to bet everything they owned on the claim that it was unwise to trust the findings of climate science.

For us to adopt a strategy that is consistent with the current scientific consensus on climate change means we must seek to restrain future growth in anthropogenic GHG emissions – enough to stabilize, as soon as possible, the level of concentrations of greenhouse gases in the atmosphere. To do so, nations collectively must prevent their GHG emissions from continuing to rise and must then begin to reduce them to zero. If we fail to satisfy these two requirements in the next thirty years or so, there is some probability that we will no longer be able to get off the Hothouse Earth pathway, no matter what we decide to do. It is very likely this path will lead to severe flooding along all coastlines and the possible abandonment of major coastal cities everywhere in the world, as well as sizeable reductions in the worldwide food supply, the widespread dieback of forests, major disruptions in marine life, and other consequences. It is very likely that such impacts will begin to be experienced well before the year 2100.

It is also very likely that, if we have embarked on this pathway to Hothouse Earth during the second half of the twenty-first century, we will find ourselves unable to alter it. Among the climate-science consensus documents are the periodic, comprehensive multi-year assessments issued by a large group of scientists assembled under the auspices of the Intergovernmental Panel on Climate Change (IPCC). In their Fifth Assessment Report (2014) we read: "Many aspects of climate change and associated impacts will continue for centuries, even if anthropogenic emissions of greenhouse gases are stopped. The risks of abrupt or irreversible changes increase as the magnitude of the warming increases."

Modern science relies on evidence-based reasoning. Major discoveries in the basic and applied sciences continue to appear regularly today, at an accelerating rate, some 250 years after a steady stream of them began to accumulate. Newer examples with practical applications include genetics, vaccines, batteries, protein folding, quantum computing, nanomaterials, and many others. These discoveries become a huge part of daily life, in communications, transportation, computer hardware and software, medicine – the list is practically endless. Like them, climate science is a product of the whole endeavour of the modern sciences, particularly chemistry, physics, and geology. It is based on the same methodologies, equations, and theories, and it shares with them the collection of evidence to validate theories and hypotheses. What else is there as a basis of belief? The IPCC reports consist of thousands of pages and thousands of references by hundreds of scientists from all over the world, and by the end of 2022 these will have been fully updated five times since the first one appeared in 1990, each one incorporating the latest results from academic research published in peer-reviewed journals. *The plain fact of the matter is that there is no competing evidence-based and comprehensive information resource of any kind that challenges the results of the IPCC reports.* To the so-called climate denialists who want to play in this intellectual arena, one can simply say: "Show me the money."

The average citizen (including the author) is no expert in these sciences. So what is belief in this case reasonably based on? The answer is that it is based on an informed *sociological* explanation, namely, understanding how science has worked over centuries of time and intensive collaborative theory-building and experimental proof. All we really need to know is how that immense structure has been designed and constructed, even if we cannot understand the underlying knowledge in terms of its intrinsic claims and correctness. If one is concerned about whether the results of climate science are reliable or not, there is no point in searching the Internet for alternative opinions (for example, that recent global warming can be explained by variations in solar radiation, despite the evidence presented in 4.1, above) – because without full training in the requisite disciplines, one cannot evaluate which of the arguments is more reliable than any of the others.

A casual expression of disbelief in such a case as this is a mug's game, a lazy search through a random assortment of contrarian opinions. There is a far more profitable venture open to anyone who is not expert in the sciences but wishes to test the basis of belief in its findings: Access a free subscription to the electronic edition of *Quanta Magazine* and read its daily reports about how scientific collaborations in recent times have been producing new results and innovative technologies, whether it is in astrophysics (the discovery of gravitational waves), or in biology (RNA vaccines or the new targeted drugs using genetics), or in chemistry (nanosensors and dual-ion batteries) – as well as the latest findings in climate science.

Scientists are trying to tell us that, starting in 1950, humanity entered a radically new phase of recent climate history. They warn us that it is very risky to continue down this path indefinitely. Why would we not believe what they are saying? Those of us alive today may think that a throw of the dice in the climate casino is a casual affair, a momentary act carried out before we turn our attention to more immediate concerns. None of our descendants, however, will be permitted to be indifferent bystanders when the results of this wager finally come in.

5

Canada: Negotiating Climate Change

Some statements made by climate scientists referenced in the previous chapters amount to stark warnings about a possibly catastrophic future for humanity, and they tell us in no uncertain terms that the nations of the world cannot just sit idly by as climate change unfolds. But what should be done, by whom, and how? Should each nation look to its own devices and decide what level of collective effort among its citizenry is required to satisfy its conscience in this matter? Or, perhaps, should everybody in any country just assess their own personal contributions to rising greenhouse-gas emissions and change their lifestyles accordingly? For example, can they just opt to have rooftop solar panels installed on the houses they own, or choose to buy electric vehicles, or just ride the bus instead, and then forget about the whole issue?

The scientific account of climate forcings and the greenhouse effect holds the three key answers to these questions. The first is the strong likelihood that human activities are now influencing the climate and thus that virtually everyone on the planet is contributing to this effect to some degree. The second is that the geophysical processes which lead to climate change are global phenomena. All the elements of climate reviewed in earlier chapters – solar radiation, the carbon cycle, the hydrological cycle, the land-atmosphere-ocean interactions – mix and circulate across the globe, no matter where they originate. The third is that human actions taken in the past, and not just present or future actions, are a major factor in climate change.

Taken together, these three answers mean that no one can sensibly make only personal choices about his or her role in global warming and not also be conscious of what others are doing, even those living in distant places: People in India have a legitimate interest in what decisions about energy use Canadians are making now and have made in the past, and Canadians have an equally vital interest in the choices made by the Chinese today.

Clearly these three answers complicate matters greatly, for if energy-use choices made by people in India, Canada, and China are all somehow interrelated, despite the distances between them all, how can they possibly coordinate their actions? Of course, options for nations in past centuries have often included going to war or, alternatively, signing multilateral treaties involving a few parties. On the eve of the First World War in 1914, competing treaties, such as the "triple entente" of Britain, France, and Russia on the one side, and the "triple alliance" of Germany, Austria-Hungary, and Italy, on the other, encouraged large-scale warfare rather than preventing it. The old system of competing power blocs aligned against each other by treaties among groups of nations broke down entirely in the twentieth century, leading to the formation of the United Nations in 1945 and the promise of agreements uniting all nations.

Modern international treaty law began about four hundred years ago in Europe, its early notable event being the Peace of Westphalia in 1648, which ended the bloody Thirty Years' War; its origins, however, can be traced back to antiquity. After 1945 it began to develop rapidly, although with significant difficulties in terms of implementation (as discussed later in this chapter). Climate change became the subject of this effort with the 1992 United Nations Framework Convention on Climate Change (UNFCCC). Treaties of this type go through the separate stages of negotiation, signing, ratification, and realization. Signing typically takes place at a meeting where general agreement on the final text of the treaty has been achieved among national representatives; ratification occurs when legal authorities in the various nations formalize their assent. Treaties often have a threshold, consisting of a specific number of states which must ratify, and when that number is reached, the treaty "comes into force," which means that its provisions are now legally or morally binding on those countries. The UNFCCC was negotiated at the "Earth Summit" in Rio de Janeiro in June 1992 and was adopted by a vote in the United Nations in that same month; it attained sufficient national ratifications to come into force in March of 1994, and at present has 197 signatories.

It has been generally agreed since 1992 that only an international treaty, signed by at least all the nations which emit large amounts of greenhouse gases, can address the threat of global warming. Therefore, the fundamental objective of all national and collective efforts made under the convention since 1992 has been to achieve "stabilization of greenhouse gas concentrations in the atmosphere at a level that would prevent dangerous anthropogenic [human-caused] interference with the climate system." *Stabilization* means the prevention of further annual increases of GHG *emissions*, from all nations of the world taken together, over some period of time sufficient to prevent *concentrations* of GHGs in the atmosphere from continuing to rise. Second, in order to prevent "dangerous anthropogenic interference with the climate system," it is desirable to reduce over time the levels of those concentrations that already exist. Finally, Article 3.1 contained an important caveat, the implications of which we will return to later:

> The Parties should protect the climate system for the benefit of present and future generations of humankind, on the basis of equity and in accordance with their common but differentiated responsibilities and respective capabilities. Accordingly, the developed country Parties should take the lead in combating climate change and the adverse effects thereof.

This provision would bedevil all subsequent attempts to implement the objectives of the convention fully and fairly.

OZONE AND CLIMATE

By 1992, Canada had become a significant player in international attempts, leading up to the creation of the UNFCCC, to reach a broad consensus among nations about what – if anything – should be done about climate change. Notably, in June 1988 the International Conference on the Changing Atmosphere: Implications for Global Security was held in Toronto; it was chaired by Stephen Lewis, then Canada's ambassador to the United Nations, and was opened by two prime ministers, Brian Mulroney of Canada and Gro Harlem Brundtland of Norway.

The Toronto conference was a direct outgrowth of an earlier effort to forge an international agreement on safeguarding the ozone layer in the earth's upper atmosphere. This process had culminated at the United Nations Vienna Conference on the Protection of the Ozone Layer in 1985 and the Montreal Protocol on Substances that Deplete the Ozone Layer in 1987. This protocol completed the task of setting a schedule for phasing out the production and consumption of such substances around the world; the fact that Montreal was chosen as the location for this important work indicated the leading role that Canada was beginning to assume in such matters of international pollution control.

There is an inherent link between the ozone-layer deliberations and the later ones on climate change. The earth's ozone layer is a very thin region of the stratosphere, which lies between twenty-five and fifty kilometres above the earth's surface and which absorbs most of the intense ultraviolet radiation (UVA and UVB) in the sun's rays. Although a small amount of UV radiation is beneficial for life on Earth, a larger amount is highly dangerous to both plants and animals, because it damages DNA and in humans leads to higher rates of skin cancer. Ozone traps something like 98 per cent of the sun's UV radiation, even though the concentration of ozone in the stratosphere is miniscule, only about ten parts per million, and for that reason the vital protection it offers to us is vulnerable to even a small amount of disruption. This vulnerability was brought to wide public attention by the phrase "the ozone hole over Antarctica," which was used in the scientific journal *Nature* in 1985.

The most dangerous of the "substances that deplete the ozone layer" are chlorofluorocarbons (CFCs) and hydrochlorofluorocarbons (HFCs). These are industrial chemicals whose primary use has been in cooling technologies, such as refrigeration and air-conditioning, and during their utilization some amounts escape and migrate into the upper atmosphere, where they destroy ozone molecules. These chemicals were soon recognized as also being potent climate-warming substances, one thousand times more powerful than carbon dioxide as heat-trapping gases, and scientists had become aware that they posed a double threat.

The fact that there was this intrinsic link between the issues of ozone protection and climate change became important in another context as well. The battle against ozone depletion was remarkable in the little time (just over five years) there was between its discovery, published in scientific journal articles, and the reaching of a consensus on both the scientific and political levels that something must be done about it. Initial opposition to abandoning CFCs from some national governments and major industry was overcome quickly, allowing international negotiation to proceed toward a resolution of the issue.

In all, only fifteen years passed between the publication of the two ground-breaking scientific articles in 1974 and the implementation of the Montreal Protocol in 1989. In large part, of course, this early success was facilitated by the fact that major industry actors had alternative solutions available in order to maintain operations of their cooling technologies while moving away from the use of ozone-depleting chemicals. This process is ongoing: the "Kigali Amendment to the Montreal Protocol" (2016), specifying targets for the reductions of HFCs, entered into force on 1 January 2019.

When the scientific community began to turn its attention from the issue of ozone to that of climate change, following the United Nations Vienna Conference in 1985, it is possible that many assumed a similar quick victory could be achieved in the global-warming case. The thought was that a similar process could be utilized: first, a UN-sponsored framework conference; second, a protocol, with targets for reducing chemical substances (in this case, greenhouse gases) in the atmosphere; and third, an enforcement regime to encourage nations around the world to reach those targets. As mentioned, Canada was at the centre of this strategy, hosting the meetings for the Montreal Protocol in 1987 and the Toronto Conference only a year later.

This expectation, however, turned out to be a fundamental mistake – with some very serious consequences. As I will explain later in more detail, in Canada and elsewhere in the world, both scientists and political leaders badly underestimated the importance of one essential difference between ozone and climate: namely, the centrality of the dependence on fossil-fuel energy dependence (coal, oil, natural gas) in national economies. Scientists and politicians indeed accomplished the first step, namely, a UN conference (UNFCCC); they finalized the next two steps with the Kyoto Protocol (adopted in 1997 and coming into force in 2005). And there, for all practical purposes, the whole process ground to a halt. Some governments and major industries began to put up a determined and prolonged campaign of resistance to reductions in the use of fossil-fuel energy, the main driver of global warming. This resistance was notable in both Canada and the United States, and, in some respects, it continues down to the present day.

THE TWENTIETH-CENTURY EXPERIENCE
WITH INTERNATIONAL TREATIES

As the discussions to follow demonstrate, in the years after 1990 the nations of the world have had great difficulties in crafting a workable treaty to deal with climate change. An all-too-common feature of international treaties is the willingness of national actors, especially the most militarily powerful ones, to renege on earlier solemn commitments, usually by simply neglecting to comply with their provisions. A sample of relevant international treaties that were negotiated during the twentieth century, starting with the Chemical Weapons Convention, illustrates this problem.

An international peace conference held in The Hague in 1899 banned the use of projectiles filled "with asphyxiating or deleterious gases." We know the ultimate result: weapons utilizing poisonous gases were used on a massive scale during the First World War, resulting in more than a hundred thousand fatalities and a million casualties. After the First World War, new poisonous substances were invented, especially nerve gases, which led to a further international ban, the 1925 Geneva Protocol for the Prohibition of the Use in War of Asphyxiating, Poisonous, or other Gases, and of Bacteriological Methods of Warfare. Nevertheless, the Italian army under Mussolini (a signatory to the protocol) used poison gas in Ethiopia in 1935–36, and at the outbreak of the Second World War there was a great deal of concern about a repeat of what had happened in Europe earlier. Large stocks of chemical weapons had been produced and held by all belligerents at the outbreak of war in 1939, though none was ever used, almost certainly because all the parties knew that any first use would prompt massive retaliation.

For the entire decade from 1961 to 1971, once again in a theatre of war far from the lands of the major powers, in Vietnam, chemical warfare was used, this time by the United States. Agent Orange, a defoliant herbicide, had extensive long-term adverse health effects associated with its contaminant, dioxin. At the same time, starting in the 1960s, an updated chemical-weapons convention was being negotiated, and it came into force in 1997. In the meantime, though the 1925 Geneva Protocol had remained in effect, Iraq under Saddam Hussein waged chemical war against Iranian troops in the 1980s and in 1991 deployed chemical weapons against some of its own citizens, the Kurdish minority.

The Biological Weapons Convention has suffered a similar fate. Biological weapons were first mentioned in the 1925 Geneva Protocol. The issue was reopened in 1968, and a Convention on the Prohibition of the Development, Production, and Stockpiling of Bacteriological (Biological) and Toxin Weapons and on Their Destruction came into force in 1975. But verification or enforcement provisions were entirely absent in the agreement and, notoriously, it was blatantly violated without penalty – especially by the Soviet Union's "Biopreparat" program.

The 2017 UN Treaty on the Prohibition of Nuclear Weapons, which "prohibits States Parties from developing, testing, producing, manufacturing, acquiring, possessing, or stockpiling nuclear weapons or other nuclear explosive devices," shows the farcical side of this business: All of the nine countries presently holding such weapons (the United States, Russia, China, the United Kingdom, France, India, Pakistan, Israel, and North Korea) have boycotted all meetings related to this treaty. The bottom line in all of this is that there are still fifteen thousand nuclear warheads in the world, almost 90 per cent of which are held by the two nuclear superpowers, the United States and Russia. The newest technologies developed by the two nuclear superpowers have resulted in the upgrading of their nuclear arsenals to truly fearsome heights, including land-based hypersonic ICBMs that travel at five times the speed of sound (faster ones are coming), each equipped with fifteen multiple independent reentry vehicles (MIRVs), which are nuclear warheads of enormous destructive power. Others are loaded onto bombers or continually roam the seas in submarines and can be fired from under water. Both countries continue to steadily refine these technologies, "improving" yields, precision in targeting, and radar-evasion techniques.

Other relevant treaties are the Convention on Biological Diversity (1993) and the Convention to Combat Desertification (1996); the Convention on the Prohibition of the Use, Stockpiling, Production and Transfer of Anti-Personnel Mines and on their Destruction (1999), with 165 signatories; the Minamata Convention on Mercury (2013), with 128 signatories and eighty-five ratifications; the International Convention on the Regulation of Whaling (1948), accepted by eighty-nine states as of 2014. All these conventions are very limited in scope and highly deficient in terms of verification and implementation. The "Whaling Convention" in particular has had a strange and controversial history, with some countries first ratifying it and then withdrawing from it more than once.

I emphasize the problematic history of international treaty negotiations on issues of global importance in order to provide some necessary context for the severe difficulties encountered in the attempts over thirty years to realize the key objectives of the UNFCCC, difficulties that, to a large extent, have still not been overcome.

THE CLIMATE-CHANGE TREATY PROCESS

As noted, this process began with the Toronto Conference on the Changing Atmosphere in 1988. It was a very high-profile international event, with over thirty senior ministers attending, in addition to prime ministers Brundtland and Mulroney. Canada's eminent academic authority, Kenneth Hare, gave the opening plenary address. The Conference Statement, a consensus view of the national representatives present, opened as follows: "Humanity is conducting an unintended, uncontrolled, globally pervasive experiment whose ultimate consequences could be second only to a global nuclear war. The Earth's atmosphere is being changed at an unprecedented rate." It continued: "The best predictions available indicate potentially severe economic and social dislocation for present and future generations, which will worsen international tensions and increase risk of conflicts among and within nations. It is imperative to act now." The echo of the famous 1957 statement by Revelle and Suess was particularly evident.

The Conference Statement explicitly pointed to a key aspect of climate change that would turn out to bedevil all those who were committed to persuading the public that it should support immediate measures to address the issue: "There can be a time lag of the order of decades between the emission of gases into the atmosphere and their full manifestation in atmospheric and biological consequences. Past emissions have already committed planet earth to a significant warming." This time lag has meant that citizens cannot readily see evidence for the slowly building risk and do not realize that they are unwittingly contributing to the emergence of future problems.

Most importantly, the statement issued a call to action applicable to all the nations of the world, which was the first exhortation of its type relating to climate change. It would be followed by many other similar calls over the next thirty-plus years:

Set energy policies to reduce the emissions of CO2 and other trace gases in order to reduce the risks of future global warming. Stabilizing the atmospheric concentrations of CO_2 is an imperative goal. It is currently estimated to require reductions of more than 50% from present emission levels.

Reduce CO_2 emissions by approximately 20% of 1988 levels by the year 2005 as an initial global goal. Clearly, the industrialized nations have a responsibility to lead the way, both through their national energy policies and their bilateral and multilateral assistance arrangements.

The reference to the need for "reductions of more than 50% from present emission levels" of greenhouse gases made little impression at the time, but it would haunt the process of climate-change negotiations thereafter. Exactly thirty years later, in 2018, new leaders at a meeting under UNFCCC auspices named the "Talanoa Dialogue" would reiterate the call for a 50 per cent reduction in GHG emissions from 2018 levels "by 2030." The first call went unheeded; the second is quite unlikely to have a better chance of being fulfilled.

The Toronto Conference parties called upon all nations to "support the work of the Intergovernmental Panel on Climate Change" (IPCC), a newly formed body that brought together specialists in climate science from around the world to provide on an ongoing basis a definitive and objective summary of what was known about the changing climate and its potential impacts on human societies. Finally, the Conference Statement asked the participants present in Toronto to "*initiate the development of a comprehensive global convention* as a framework for protocols on the protection of the atmosphere." The uptake on this last expectation was quick indeed. Only four years later, in May-June 1992, the United Nations Conference on Environment and Development, held in Rio de Janeiro and known informally as the "Earth Summit," produced the United Nations Framework Convention on Climate Change (UNFCCC). Once again, Canada played a leading role: Prime Minister Brian Mulroney attended the conference and signed the document, and later Canada was the first G7 nation to ratify the convention.

Much of the discussion to follow here will focus on the setting of targets for the reduction of GHG emissions. The Toronto Conference initiated this process in terms of three stipulations: (1) to "stabilize" the atmospheric concentrations of CO_2; (2) to reduce current global levels of those emissions by 50 per cent; and (3) to reduce 1988 emissions by 20 per cent by 2005 as an initial step. "Stabilizing" GHG emissions means to get them to stop rising, permanently. Objective number two had no timeline, but number three gave a firm interim target. Canada's Mulroney government signed on to these commitments.

When the text of the UNFCCC document was finalized a few years later, it contained (as noted) a new general objective for the nations collectively: to seek to prevent "dangerous anthropogenic interference with the climate system." This has remained the foundation stone for all subsequent action plans down to the present day, but, in the articles of the convention itself, no specific policy measures to fulfill that objective were offered, nor was there any guidance as to how great an effort might be needed in order to reach this goal. But in a non-binding addendum to the basic convention articles, industrialized countries including Canada agreed to a newer commitment, the second in a long series: to return to their 1990 levels of GHG emissions by the year 2000. Countries went home to consider their options for meeting these commitments – and that's when the trouble started. This trouble was compounded by the replacement of the Conservatives by the Liberal Party in the federal election of October 1993.

It is a lamentable truth that, for the following five years, Canadian governments, both federal and provincial, tossed back and forth proposals for GHG reductions targets through a bewildering series of policy proposals while actually achieving absolutely nothing. Worse, the country's GHG emissions were steadily rising during this period, making the achievement of any reductions progressively harder. The whole sad story is well told in the 2007 book *Hot Air*, written by Jeffrey Simpson, Mark Jaccard, and Nic Rivers. This account is still very much worth reading today, because the temporizing, long-discredited rationales and endless excuses for inaction described therein are still alive and well in Canada.

One episode in particular stands out, since it had a major impact on the following decade of government inaction in this country. An international meeting held pursuant to the UNFCCC is called a "conference of the parties" (COP); the first one was held in Berlin in 1995 and the most recent (COP26) in Glasgow in late 2021. The 1995 meeting and the following one were to prepare for negotiations on actual numerical targets for emissions reductions, applicable to only a selected list of developed nations – thirty-nine in all, including Canada and the United States, which would become the so-called "Annex-B" nations. These targets would be debated at COP3, to be held in December 1997 in Kyoto, Japan. In advance of the departure of the Canadian team of negotiators sent to Kyoto, there was a federal-provincial meeting held in Regina, Saskatchewan, where general agreement was reached supporting a ten-year extension of the 1992 UNFCCC target (i.e., reduction of emissions to 1990 levels). However, instead of extending this 0 per cent/1990 target to year 2010, Canadian officials, once having arrived in Kyoto, opted for a –6 per cent/1990 national commitment in order to try to "beat" the initial –5 per cent/1990 target set earlier by the United States negotiators. This turned out to be not only a betrayal of the federal/provincial meeting but also a fruitless strategy: at Kyoto, the United States agreed to a –7 per cent/1990 target and then failed to ratify the Protocol! The oil-producing provinces in Western Canada were understandably outraged. The government of Prime Minister Jean Chrétien took five years (until 2002) just to ratify the Kyoto Protocol, worse, both his government and the successor

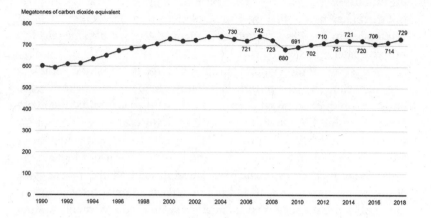

Figure 5.1 | Canada greenhouse gas emissions.

Liberal government, under Paul Martin at this point but in power as a party from 1993 to 2006, made no real effort during all this time to meet any such target.

The charade began anew under the government of Stephen Harper's Conservative Party. As *Hot Air* explains:

> The [2007 Harper] government pledged to reduce Canada's total GHG emissions to 20 percent below their 2006 levels by 2020, en route to a 50 to 70 percent reduction by 2050 ... Remember that Canada under the Chrétien liberals had committed to reducing Canada's GHG emissions by 6 per cent from 1990 levels by 2008–12. Instead, by 2004 they had risen by 26 per cent, [missing our] Kyoto target by 33 per cent.

Neither this nor any other government under Prime Minister Stephen Harper between the years 2006 to 2015 ever met any GHG emissions-reductions targets whatsoever. The Conservative Party led by Harper had a strong Alberta contingent, among whom the residual ill feeling from what happened in 1997 remained strong. This and other reasons eventually resulted in Canada withdrawing from the Kyoto Protocol in 2012, the only country ever to do so. Under the Doha Amendment to the Protocol in 2012, various countries made new commitments for greenhouse-gas emissions for the period 2013 to 2020. But Canada was not there, having notified the United Nations on 15 December 2011 that it intended to withdraw from the Protocol, a step that became effective one year later.

The position of the United States looms large in this story. Although President Clinton had committed his country in 1997 to a −7 per cent/1990 emissions-reduction target at Kyoto, he never even submitted the Kyoto Protocol to the US Senate for ratification (which is a requirement for all treaties under the US Constitution), because he knew it would never pass. The United States had ratified the UNFCCC earlier, but already in 1997 the Senate had passed the Byrd-Hagel Resolution by a vote of 95–0. It stated that the country would not be a party to any protocol negotiated under the UNFCCC that mandated (1) "new commitments to limit or reduce greenhouse gas emissions [for developed nations], unless the protocol or other agreement also mandates new specific scheduled commitments to limit or reduce greenhouse gas emissions for Developing Country Parties within the same compliance period," or (2) that "would result in serious harm to the economy of the United States." In other words, President Clinton sent his negotiators to Kyoto advertising a very ambitious emissions reductions target that he knew in advance would never be fulfilled.

Representatives of the United States nonetheless continued to participate in COP meetings, and then President Obama signed the 2015 Paris Agreement, which was designed as an "action" under the UNFCCC with non-binding commitments. Therefore, it did not require either ratification by the Senate or any other type of formal acceptance. Nineteen years after the Byrd-Hagel Resolution, in the aftermath of the Paris Agreement (discussed in the next chapter), the US Senate was again strongly critical of Obama's new effort. Then, in November 2017, President Trump announced his intention to withdraw from the Paris Agreement, although, under the terms of the agreement, the United States could not issue a formal notice of withdrawal until November 2019 and could then only actually leave the agreement on 4 November 2020. President Trump did follow through on these steps, and the United States exited the Paris Agreement exactly one day after the presidential election held on 3 November. As he had promised to do, Joe Biden rejoined the agreement on the day of his inauguration as president on 20 January 2021. As one can easily imagine, this flip-flopping by the world's second-largest emitter of greenhouse gases did nothing to bolster confidence in the international treaty regime.

What about the performance of the rest of the "Annex-B" developed countries that signed and then ratified the Kyoto Protocol, which committed them to making specific reductions in GHGs between 2008 and 2012? An analysis was undertaken by Michael Grubb, who found that, leaving aside the United States and Canada, the remaining Annex-B countries with targets, taken as a whole, achieved at least a "kind of" 100 per cent compliance with their commitments, notably by relying on Kyoto's flexibility mechanisms:

> The collective task of the remaining 36 countries with quantified commitments – the "Annex-B-2012" countries – was very lop-sided, with potential shortfalls in many of the OECD countries alongside vast surplus in the post-Soviet era "countries with Economies in Transition (EiTs)," led by Russia and Ukraine ... Compliance did rely on project credits: about 300 MtCO2 annually during the period was secured from certified emission reduction projects in developing countries through the Clean Development Mechanism (CDM), and about half as much again came from similar project-based credits generated within the Annex-B-2012 countries ("Joint Implementation").

Here Grubb refers to the mechanisms provided in the protocol to assist developed nations in meeting their targets by claiming credits for projects financed in those that are developing. This practice has been often criticized, but in truth it represented a clear realization by the negotiators at Kyoto that the first phases of implementation would be difficult. Essentially, it conceded that getting started would be onerous, but that, once started, later stages of continuing emissions-reductions could become progressively easier, since over time the ongoing adjustments would have been built into newer phases of economic growth and administrative structures.

If there is only one lesson to be learned by Canadians with respect to our promises made in the context of joining international treaties, it is this: policies formulated in order to realize those promises should begin to be implemented the moment that our negotiators set foot in Canada upon returning home from abroad. Abundant experience tells us that anything else results in embarrassment. When Brian Mulroney made Canada's very first emissions-reduction commitment, at the Toronto Conference in 1988, our annual GHG emissions stood at 588 megatonnes (Mt), and the promise made was to reduce them by 20 per cent (that is, to 470 Mt) by 2005, which meant an average of about seventeen Mt per year. In 2002, when Canada ratified the Kyoto Protocol – committing to reducing GHG emissions by 6 per cent below 1990 levels by 2012 – Canada's emissions had grown from 1990 levels of 590 Mt to 717 Mt. Thus in 2002 we were almost 35 per cent above our original 1988 target. As we shall see in the next chapter, Canada's proclivity for failing to deliver on its promises related to international treaties on GHG reductions has continued right down to the present day.

In 2009, as the end of the Kyoto Protocol's first commitment period (2008 to 2012) was approaching, the parties to UNFCCC met at COP15 to consider what might follow Kyoto. The non-binding agreement reached there, known as the Copenhagen Accord, was the first to stipulate a specific target for controlling GHG emissions in terms of global temperature. Referring to the objective of preventing "dangerous anthropogenic interference with the climate system," the parties agreed with "the scientific view that the increase in global temperature [over pre-industrial levels] should be below 2 degrees Celsius." Then, as noted, the Doha Amendment to the protocol designed a second commitment period: "Parties committed to reduce GHG emissions by at least 18 percent below 1990 levels in the eight-year period from 2013 to 2020."

However, as of 2012, only a handful of countries plus the European Union had agreed to further GHG reductions. The later total of 126 countries which have "deposited their instrument of acceptance" for the Doha Amendment as of February 2019 do not include the two largest global emitters, China and the United States. (China accepted only for its two peripheral provinces, Hong Kong and Macao.) In addition, of the 126 signatories, a mere twenty-nine of them – the twenty-eight of the pre-Brexit EU plus Iceland – had formally agreed to fulfill their commitments under the Doha Amendment. Again, this was a kind of theatrical exercise with no real substance.

Although the terms of the Kyoto Protocol had been signed by representatives of the developed nations at the end of 1997, the protocol still needed to be ratified, and it did not actually come into force until February 2005, a full eight years later. "Still-developing" nations had no commitments under the protocol, but only a year later a momentous event occurred on the climate-change front: China passed the United States as the world's leading emitter of greenhouse gases, and in subsequent years China's emissions skyrocketed, mostly as a result of coal combustion. Even as the developed world worked under the terms of the Kyoto Protocol through 2012, this drastic change in the global profile of GHG emissions made it obvious that a completely different type of international agreement was needed. This new instrument would be the Paris Agreement of 2015, which meant that, for the first time, all nations of the world would come under a single umbrella in the effort to control greenhouse-gas emissions. The Paris Agreement entered into force in 2016, and as of the time of writing in 2022, it is still the framework under which this effort continues to be pursued.

The latest specific target set by the Secretary-General of the United Nations is for the world to achieve "net zero" GHG emissions (also called "carbon neutrality") by 2050. This means that a country either emits no GHGs or any amount it does emit is offset by other measures, such as by tree-planting or capturing carbon from the air. As of now, over a hundred countries have promised to try to reach this 2050 goal; some, including France, the United Kingdom, Sweden, New Zealand, and Canada, have enshrined this target in legislation. China, the world's largest emitter, has promised to reach net zero "before 2060." Let us review Canada's situation in recent years to see whether it is likely that it can meet this test.

Canada's original pledge to the world under the Paris Agreement was that we would reduce GHG emissions by 30 per cent below 2005 levels by 2030. Our emissions in 2005 were 730 Mt, and thus our 2030 target was 511 Mt. (Note that Canada's emissions as of 2019 were virtually the same as in 2000, as shown in figure 5.1, despite the many pledges for reductions made in the interim.)

In 2018, a group comprising all of Canada's auditors-general, provincial and federal, issued a comprehensive report titled *Perspectives on Climate Change Action in Canada*. The report noted that Canada had missed the targets it had set for itself in 1992 in Rio (stabilizing at 1990 levels by 2000) and the Kyoto target in 1997 (6 per cent below 1990 by 2012), and very likely would miss a new target (17 per cent below 2005 by 2020) accepted at the COP meetings in 2009 and 2010. It continues:

> Canada's auditors general found that most governments in Canada were not on track to meet their commitments to reducing greenhouse gas emissions and were not ready for the impacts of a changing climate. On the basis of current federal, provincial, and territorial policies and actions, Canada is not expected to meet its 2020 target for reducing greenhouse-gas emissions. Meeting Canada's 2030 target will require substantial effort and actions

beyond those currently planned or in place. Most Canadian governments have not assessed and, therefore, do not fully understand what risks they face and what actions they should take to adapt to a changing climate.

The report did neglect to mention the initial pledge – made by Prime Minister Mulroney at the Toronto Conference – of a 20 per cent reduction from 1988 levels by 2005.

The website Climate Action Tracker, which is sponsored by a consortium of non-profit groups affiliated with Germany's Potsdam Institute for Climate Impact Research, has been tabulating pledges and actions worldwide, on a country-by-country basis, for over a decade. The site states: "While emissions in 2020 are projected to fall by 11 to 13% compared to 2019 due to the economic impact of the pandemic, Canada is still likely to miss its 2030 NDC by 15–20% as we estimate that emissions could be in the 603–630 MtCO$_2$e range." It also argues that Canada's climate commitment in 2030 is "not consistent with holding warming to below 2°C." We will all have to wait and see what actually happens. For some years, a complicating factor had been a lawsuit launched by several provincial governments against the Government of Canada, claiming that it had no authority to impose a nationwide carbon tax. In 2016, a careful analysis by legal scholar Nathalie Chalifour was published, arguing that the federal power in this area is unequivocal; in March 2021 the Supreme Court of Canada upheld the federal government's Greenhouse Gas Pollution Pricing Act by a 6–3 vote.

A NEW LAW AND A NEW TARGET IN 2021

In November 2020, the Government of Canada for the first time did put a future target – "the national greenhouse gas emissions target for 2050 is net-zero emissions" – into a proposed law in Bill C-12: Canadian Net-Zero Emissions Accountability Act, which received royal assent on 30 June 2021. The text of the law stipulates that the minister of environment and climate change would "establish an emissions reduction plan for 2030 within six months after the day on which this Act comes into force." Furthermore, there is an interim greenhouse gas objective for 2026, as well as three progress reports to be submitted no later than the end of 2023, 2025, and 2027. The minister is required to set 2035, 2040, and 2045 targets at least ten years in advance. Finally, the act tasks the commissioner of the environment and sustainable development to examine and report on the Government of Canada's implementation of measures aimed at mitigating climate change at least once every five years, starting no later than the end of 2024.

At a climate summit on Earth Day (April 22) in 2021, US president Joe Biden almost doubled that country's initial emissions-reduction target under the Paris Agreement, which had been set at –27 per cent/2005 by 2025, changing it to "up to" –52 per cent/2005 by 2030. Prime Minister Justin Trudeau responded by increasing Canada's 2030 target from –30 per cent/2005 to between –40 per cent and –45 per cent/2005. Meeting the original target put Canada's emissions in 2030 at 511 Mt; assuming that the midpoint of the new target is achieved (–42.5 per cent), emissions in 2030 will be 420 Mt.

There are some serious risks for Canada in seeking to match US ambitions in this domain, just as there were more than two decades ago in 1997 at Kyoto (see the discussion earlier in this chapter). During the period from 1990 to 2019, Canada's GHG emissions increased by 21.4 per cent; its per-capita emissions fell by about 11 per cent; and its emissions intensity, its GHGs per unit of GDP, fell by 37 per cent (Canada, *Greenhouse Gas Emissions* [2021], figures 1 and 2). Here are the comparable numbers for the United States for the same period: (1) total GHG emissions, +1.8 per cent; (2) per capita, –22 per cent; (3) intensity, –50 per cent (EPA, *Inventory* [2021], 2–41). Thus the United States will be entering the new target arena with an economy that has driven down per-capita and emissions-intensity numbers over the past three decades, even in the absence of political support at the national level; and the Biden administration will now provide a powerful additional impetus.

In the "Mitigation" section of chapter 8, below, I argue that Canada will have to struggle to maintain and achieve even its original 2030 target (–30 per cent/2005), much less its enhanced one (–40 per cent to –45 per cent/2005), throughout the decade of the 2020s. For example, as some commentators mentioned on 22 April, Canada's electricity sector has already been largely decarbonized (unlike in the United States), and emissions are still rising in the oil and gas sector. Surely, delivering on the new target promise is likely to be a great deal harder. In this context, however, the commitment to a series of interim targets, to frequent progress reports, to decade-long lead times for later targets, and to regular oversight reporting by the commissioner of the environment and sustainable development are all now, for the first time, enshrined in legislation. This will make it much more politically difficult for later governments to ignore or adjust targets.

The most important new target, however, is that of 2050. As of spring 2022, seventy countries (the EU plus thirty-three others) had set a "carbon neutrality" target – that is, net-zero carbon emissions – by that year. The International Energy Agency, introducing its May 2021 report (IEA 2021b), *Net Zero by 2050: A Roadmap for the Global Energy Sector*, had this to say about reaching the 2050 target:

> Doing so requires nothing short of a total transformation of the energy systems that underpin our economies ... Despite the current gap between rhetoric and reality on emissions, our Roadmap shows that there are still pathways to reach net zero by 2050. The one on which we focus is in our analysis the most technically feasible, cost-effective and socially acceptable. Even so, that pathway remains narrow and extremely challenging, requiring all stakeholders – governments, businesses, investors and citizens – to take action this year and every year after so that the goal does not slip out of reach.

This message was reinforced by economist Jean Pisani-Ferry in an important essay published online in August 2021, emphasizing the large macroeconomic impacts of the net-zero pledge. He warned that "decades of procrastination" will make the transition to a decarbonized future much more difficult than it might have been – and that any further procrastination may put that objective realistically out of reach. Canada and every other nation that has made the 2050 pledge would do well to take this advice to heart.

Commenting on the IEA report, the University of Ottawa's Monica Gattinger writes: "This constitutes a complete remaking of energy and economic systems over the next 30 years." She also recognizes that the IEA's position that "there should be no development of new oil and gas fields if the world is to achieve net zero targets" represents a serious political challenge for a country such as Canada. Added to this are the technology development imperatives in terms of batteries, heavy-industry emissions, a huge expansion of the electricity grid and CCS (carbon capture and storage) facilities, blue hydrogen production, new nuclear and renewable energy installations, and other initiatives. By 2035, governments would have to ban further production of internal-combustion engines for personal vehicles and for half of the heavy truck fleet. The list goes on. As University of Cambridge political economist Helen Thompson puts it, "transitioning away from fossil fuels and toward greener energy ... requires nothing less than changing the material basis of modern civilization."

The great value of the IEA report is that it shows so clearly the real face of the net-zero-2050 pledge in terms of practical impacts on economy, lifestyles, jobs, technologies, and public policy. The IEA report's perspective has been reinforced by a major Canadian study, *Canadian Energy Outlook 2021*, released in October 2021 by Montreal's Trottier Research Institute, which also carefully examines, sector by sector, the serious difficulties that Canada will face in seeking to meet its current emissions-reductions targets for 2030 and 2050. In another report released in October 2021, the Royal Bank of Canada (RBC 2021) estimated that $2 trillion of new investments over the next thirty years – an average of $60 billion per year – will be needed for Canada to reach net-zero GHG emissions by 2050.

It may be no exaggeration to say that fulfilling these pledges for Canada will require a collective level of effort last seen during the Second World War.

6

Treaty Framing and Climate Science

In this chapter we confront the yawning gap – for the world as a whole – between necessary GHG emissions-reductions targets proposed by climate scientists, on the one hand, and the results of decades-long international treaty negotiations and meetings, on the other. Thirty years after the treaty process commenced, this gap is still growing. Here the process of international treaty negotiation will be juxtaposed more systematically with the concept of *climate forcing* that is the mainstay of the scientific account of global climate change. On the other hand, I define *treaty forcing* as the variety of processes whereby all or some of the nations that emit greenhouse gases (GHGs) use an international treaty framework to determine how and when emissions reductions should occur.

To summarize our earlier discussion, climate forcing – also known as radiative forcing – can be defined as an imposed perturbation of Earth's energy balance. Energy flows in from the sun, much of it in the visible wavelengths, and back out again as long-wave infrared (heat) radiation. An increase in the luminosity of the sun, for example, is a positive forcing that tends to make Earth warmer. A very large volcanic eruption, on the other hand, is a negative forcing, since it increases the aerosols (fine particles) in the lower stratosphere at altitudes of sixteen to twenty kilometres, and these reflect sunlight back into space and thus reduce the solar energy delivered to Earth's surface. These examples are natural forcings.

On the other hand, human-caused forcings result from such things as the gases and aerosols produced by burning fossil fuels and alterations to the earth's surface from various changes in land use, such as the conversion of forests into agricultural land. Those gases that absorb infrared radiation, that is, the "greenhouse" gases, tend to prevent this heat radiation from escaping to space, leading eventually to a warming of the earth's surface. The observations of human-induced forcings underlie the current concerns about climate change.

International environmental treaty-law processes seek voluntary acceptance by nations of limitations on those of their practices that have an appreciable impact on global conditions. The major actions to date are the United Nations Framework Convention on Climate Change (1992) and its following steps, the 1997 Kyoto Protocol and its Doha Amendment, and the 2015 Paris Agreement. It was Article 2 of the UNFCCC that characterized the ultimate objective of global action as "stabilization of greenhouse gas concentrations in the atmosphere at a level that would prevent dangerous anthropogenic interference with the climate system." In attempting to meet this objective, subsequent developments have moved from a largely top-down strategy (Kyoto and Doha), under which specific numerical emissions-reductions targets were set for some nations, to a bottom-up one (Paris), a more inclusive setup than its predecessors, under which all nations identify their own targets in terms of reductions or alternative measures.

It is important to know that there are other contributors to climate forcing that are not covered in the prevailing international treaties on climate change. These contributors include land use and land-use change, already mentioned, and black carbon, which according to some estimates vies with methane as the second-largest factor in climate forcing after carbon dioxide. Black carbon, and to a lesser extent brown carbon, are short-lived particulate emissions, not gases, which are generated by the incomplete combustion of fossil fuels and biomass (renewable organic material from plants and animals), and which have a major impact on human health as well as on climate change. Advanced industrial economies have taken important steps to reduce emissions of black carbon and are committed to achieving further reductions.

With few exceptions across the world, GHG emissions in 2020 will show a sharp year-over-year reduction caused by the pandemic. Very likely it will take until early 2023, when the 2022 numbers are in following the first full year after the coronavirus is expected to be largely brought under control in the developed economies, including China, before we will be able to ascertain whether longer-terms progress in setting global GHG emissions firmly on a path toward stabilization is finally being made. However, in early 2021, the International Energy Agency (IEA) issued its projections for both 2021 and 2023 in its *Global Energy Review*. Whereas global CO_2 emissions had decreased by 5.8 per cent in

2020 due to the pandemic, they are projected to increase again by 4.8 per cent in 2021, thus recovering 80 per cent of the earlier reduction in one year (IEA 2021a).

In October 2021, IEA produced the first report of its "Sustainability Recovery Tracker" (IEA 2021c) designed to monitor energy-related government policies and both public and private spending on clean energy and to project the effect of these on future CO_2 emissions. This first report states: "We estimate that full and timely implementation of the economic recovery measures announced to date would result in CO_2 emissions climbing to record levels in 2023, continuing to rise thereafter."

FROM KYOTO TO PARIS

In this section, the twin stories about climate forcing and treaty forcing are brought up to date for the period beginning in 2015. As we have seen, around 1990 the nations of the world had embarked on a sustained, collective effort to limit and reduce global greenhouse-gas (GHG) emissions from anthropogenic sources, an effort which continues to this day through the Paris Agreement of 2015. However, between 1990 and 2018, humanity's releases of carbon dioxide (CO_2), which account for about three-quarters of all GHG emissions, increased by 67 per cent. In fact, according to the US National Oceanic and Atmospheric Administration, the growth rate of CO_2 emissions is accelerating, having been substantially higher in the most recent decade (2010 to 2019) than it was in the two prior decades. Peters et al. (2019) sum up as follows: "The continued growth in global fossil CO_2 emissions is taking place despite growing public and policy attention, five cycles of IPCC Assessment Reports and almost 30 years of international climate negotiations."

Global GHG emissions have three major components: (1) CO_2, in which combustion of fossil fuels is the primary source; (2) non-CO_2 gases (methane [CH_4] and others); and (3), a separate category for GHG emissions resulting from land use and land-use change in agriculture and changes in the earth's forest cover (LULUCF). Quantities are expressed in terms of either megatonnes (Mt) or gigatonnes (Gt). Emissions numbers for the second and third categories are normally converted into their equivalent in CO_2, abbreviated "CO_2eq" or "$MtCO_2e$/year," because the other gases, such as methane, have a different heat-trapping potential in the atmosphere. (For example, over a hundred-year period, on a per-unit basis, methane has twenty-five to thirty times the warming impact of CO_2.)

CO_2 emissions are the largest source of global greenhouse-gas emissions, with a share of about 73 per cent, followed by the non-CO_2 gases: methane (CH_4: 18 per cent), nitrous oxide (N_2O: 6 per cent), and fluorinated gases (3 per cent). The emissions numbers cited in this chapter will vary, depending on whether they are depicting GHG emissions from all sources or only fossil-fuel emissions (E_{FF}). The latest full estimate for global emissions in the pre-pandemic period is for 2019 (Friedlingstein et al. 2020):

> Preliminary estimates of global fossil CO2 emissions are for growth of only 0.1% between 2018 and 2019 to remain at 9.7 ± 0.5 GtC in 2019 … In 2019, the largest absolute contributions to global fossil CO2 emissions were from China (28%), the USA (14%), the EU (27 member states; 8%), and India (7%). These four regions account for 57% of global CO2 emissions, while the rest of the world contributed 43%, which includes aviation and marine bunker fuels (3.5% of the total). Growth rates for these countries from 2018 to 2019 were +2.2% (China), −2.6% (USA), −4.5% (EU27), and +1.0% (India), with +1.8% for the rest of the world.

Early in 2017, when the GHG emissions data for 2016 had been finalized, it looked as if this powerful driver of climate change had finally peaked, for 2016 was the third consecutive year in which global emissions had remained basically unchanged, leading to the possibility that the initial fundamental objective of climate-change action plans, namely GHG stabilization, had been secured. But this has not been sustained. Global fossil-fuel emissions began rising again, with increases of 1.3 per cent in 2017, 2.7 per cent for 2018, and 0.1 per cent in 2019. Although the coronavirus pandemic has depressed earlier projections for rising emissions in 2020, authoritative references cited above (IEA), as well as further on in this chapter, suggest that global emissions are indeed expected to resume additional growth as of 2023 and continue rising until 2030.

Most CO_2 has an atmospheric lifetime between twenty and two hundred years, but some of it remains for thousands of years, which means that a declining proportion of the gas, once emitted, remains in the air for a very long time. Other gases have a stronger heat-trapping effect but shorter atmospheric lifetimes, for example nitrous oxide (average of 120 years) and methane (twelve years). The ozone-depleting chemicals have the highest global-warming potential of all, and some of them have very long atmospheric lifetimes. Accumulated emissions load ever more momentum into the climate-forcing mechanism, as the climate system slowly seeks a new state of equilibrium.

The long lifetime of CO_2 in the atmosphere means that emissions over the past century and even beyond, and not only recent emissions, are important. Although China is presently the largest source of current emissions, the highest *cumulative* contributors to global emissions are the United States (25 per cent), the EU-28 (22 per cent), China (13 per cent), Russia (7 per cent), Japan (4 per cent), and India (3 per cent). Thus, as of 2017, the United States, the European Union, and Japan combined accounted for 51 per cent of cumulative emissions. The attentive reader will want to explore Ritchie and Roser (2020), "Cumulative CO_2 Emissions by World Region," which is an interactive data visualization graphic depicting the estimated shares of historical emissions by year from 1751 to 2020.

TREATY FRAMING IN THE CONTEXT OF CLIMATE FORCING

As noted earlier, the UNFCCC, which came into force in 1994, set in motion what would become a long search for acceptable national targets for reducing greenhouse-gas emissions. Under the UNFCCC, the meetings in Kyoto in 1997 comprised the third Conference of the Parties (COP3).

The Kyoto Protocol came into force in early 2005, setting defined national targets – applicable only to developed countries and some "economies in transition" – for the period 2008 to 2012 in order to achieve GHG emissions reductions of an average of 5.2 per cent below 1990 levels. The detailed rules for the implementation of the Kyoto Protocol had been adopted in 2001 at COP7 in Marrakesh, Morocco (the "Marrakesh Agreements"). However, even while the UNFCCC had continued with negotiating the Bali Action Plan 2007 (COP13), the stage had been set for a fundamental reframing of the situation. A summary of the first phase is in table 6.1.

Just after the Kyoto Protocol's first commitment period had opened in 2008, a dramatic set of new ultimate objectives for emissions reductions were first broached during the negotiations on what would become the Copenhagen Accord of 2009 at COP15. "Dangerous anthropogenic interference with the climate system" was now defined as exceeding two global temperature-related ceilings of 2°C and ideally no more than 1.5°C above pre-industrial levels. One factor in the lead-up to Copenhagen

Table 6.1 | Early global GHG emissions targets

1992 UNFCCC	"Developed country parties and other parties included in Annex I" will undertake "policies and measures ... with the aim of returning individually or jointly to their 1990 levels" of "anthropogenic emissions of carbon dioxide and other greenhouse gases not covered by the Montreal Protocol"
1997 Kyoto Protocol, First Commitment Period (2008–2012), Annex I Countries	Annex I countries are industrialized countries plus "economies in transition": 39 parties including the EC (30 of which are in Europe), collective average target of 5.2% below 1990 levels, range of country targets from minus-8 to plus-10.
2007 IPCC Fourth Assessment Report, *Mitigation of Climate Change*, 776	Scenario: GHG Concentrations of 450ppm CO2eq: Annex I Countries: (a) –25% to –40% from baseline 1990 levels by 2020; (b) –80% to –95% by 2050; Annex II: "Substantial deviation from baseline" by 2020 and 2050.
2009 Copenhagen Accord	The Parties recognize "the scientific view that the increase in global temperature should be below 2 degrees Celsius," therefore: "Annex I Parties commit to implement individually or jointly the quantified economy-wide emissions targets for 2020" and "Non-Annex I Parties to the Convention will implement mitigation actions." The Accord gives no specific targets.
2012 Kyoto Protocol, Doha Amendment	Establishes a Second Commitment Period (2013–2020) with new targets for 2020, mostly for European countries, of between –20% to –30%, base year 1990.

was pressure from the Alliance of Small Island States (AOSIS), which had first raised the issue of a temperature-based limit of warming to 1.5°C in 2008; to some extent, therefore, "avoiding +1.5°C" could be described as a well-intended gesture to AOSIS that few believed had any chance of realization. On 21 December 2012, the Doha Amendment was circulated at a Conference of the Parties (COP18); this provided a second commitment period, during which parties would commit to reducing GHG emissions by at least 18 per cent below 1990 levels in the eight-year period from 2013 to 2020. As indicated earlier, the two commitment periods under the Kyoto Protocol did not achieve their objectives.

The Intergovernmental Panel on Climate Change's Fifth Assessment Report (AR5: 2014) had focused much attention on a temperature-defined objective for emissions reductions, using a set of four projections called "representative concentration pathways" (RCPs), which are scenarios of possible future developments for climate change. In AR5, the IPCC had also looked much further out in time, initially advising that an ultimate target of zero anthropogenic emissions would have to be achieved by the end of the twenty-first century: "Emissions scenarios leading to CO_2-equivalent concentrations in 2100 of about 450 ppm or lower are likely to maintain warming below 2°C in the 21st century relative to pre-industrial levels. These scenarios are characterized by 40 to 70% global anthropogenic GHG emissions reductions by 2050 compared to 2010, and emissions levels near zero or below in 2100."

Thus, already by 2014 climate scientists had concluded that the earlier basis on which the 2°C target was built was no longer valid. Given steadily rising emissions, increases in human population, and other factors in the period following 2009, many had already concluded that there were few prospects for limiting warming to 2°C above pre-industrial levels. These conclusions had been reached over the five years since the Copenhagen Accord had first placed rising global temperatures at the centre of emissions-reductions scenarios, setting the stage for an effort to insert this criterion into a renewed initiative under UNFCCC, which would seek to align new targets for emissions reductions with their corresponding temperature projections.

THE PARIS AGREEMENT OF 2015

The 2015 Paris Climate Agreement seeks to promote the eventual implementation of the UNFCCC objectives and to strengthen the global response to the threat of climate change by "[h]olding increases in the

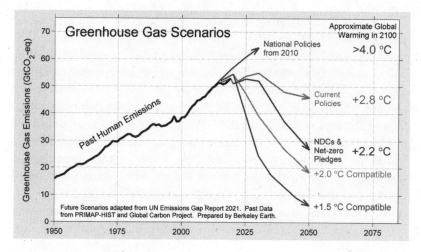

Figure 6.1 | Warming scenarios.

global average temperature to well below 2°C above pre-industrial levels and pursuing efforts to limit the temperature increase to 1.5°C above pre-industrial levels, recognizing that this would significantly reduce the risks and impacts of climate change" (Article 2.1). To achieve these temperature goals, the Paris Agreement states (Article 4.1) that parties will "aim to reach global peaking of greenhouse gas emissions as soon as possible, recognizing that peaking will take longer for developing country Parties, and to undertake rapid reductions thereafter in accordance with best available science, so as to achieve a balance between anthropogenic emissions by sources and removals by sinks of greenhouse gases in the second half of this century."

Innovations in the Paris Agreement include an entirely new framework for climate-change issues based on international governance, such as "recognizing the importance of the engagements of all levels of government *and various actors*" (Preamble, emphasis added). Article 4(b) acknowledged that some parties may "incentivize and facilitate participation in the mitigation of greenhouse gas emissions by public and private entities authorized by a Party." The agreement requires all parties to put forward their best efforts through successive voluntary commitments, called "nationally determined contributions," or NDCs, and encourages parties to strive for their "highest possible ambition" by enhancing their NDCs at any time. In October 2016, the threshold for entry into force of the Paris Agreement was achieved, and this occurred on 4 November 2016.

As we have seen, during the period following the ratification of the Kyoto Protocol in 2005, the gap between treaty-determined emissions-reductions targets and actual global emissions steadily widened. The protocol had provided that "where the enforcement branch has determined that the emissions of a Party have exceeded its assigned amount, it must declare that that Party is in non-compliance and require the Party to make up the difference between its emissions and its assigned amount during the second commitment period, plus an additional deduction of 30%." Since the second commitment period was organized under the Doha Amendment, which itself has failed to come into force, the sanctions for non-compliance under the protocol became irrelevant. This issue was sidestepped when the top-down Kyoto approach was replaced by the bottom-up approach adopted in the Paris Agreement, which allows parties to set their own targets and to decide whether they are able to meet them, and when.

The core of the Paris Agreement is the submission by the parties of their "nationally determined contributions," representing their voluntary and individually framed targets for GHG emissions reductions. Some parties had submitted "intended" or initial targets (INDCs) prior to the commencement of the Paris meetings. However, even as the Paris Agreement was being adopted at COP21, the UNFCCC Secretariat was circulating a Synthesis Report on the Aggregate Effect of the Intended Nationally Determined Contributions, which warned: "The estimated aggregate annual global emission levels resulting from the implementation of the INDCs do not fall within least-cost 2°C scenarios by 2025 and 2030 … Therefore, much greater emission reductions effort than those associated with the INDCs will be required in the period after 2025 and 2030 to hold the temperature rise below 2°C above pre-industrial levels."

At the same time, expert consensus documents, especially those issued by the Intergovernmental Panel on Climate Change (IPCC), have been seeking to be more and more precise on what emissions-reductions targets are needed in order to avoid exceeding the thresholds of 1.5°C and 2°C warming. The IPCC 2018 Special Report (SR-15), *Global Warming of 1.5°C*, states: "In model pathways with no or limited overshoot of 1.5°C, global net anthropogenic CO_2 emissions decline by about 45% from 2010 levels by 2030 … reaching net zero around 2050." It is a clear sign of the emerging dilemmas in the policy responses to climate change that, *a mere four years after AR5 had appeared, nations were accepting the need to arrive at emissions "near zero" by 2050 rather than 2100.*

Table 6.2 | INDCs for the top sixteen CO_2 emitters in Mt (Paris Agreement)

A	B Country or region	C 2019 CO_2 emissions (Mt)	D 2019 increase/ decrease	E Emissions per capita (I)[1]	F First NDC[2]
1	China	11,535 (30.3%)	+3.4%	8.1	Emissions peaking "around" 2030
2	USA	5,107 (13.4%)	–2.6%	15.5	–26% to –28% of 2005 by 2025
3	EU27+UK	3,304 (8.7%)	–3.8%	6.5	–40% of 1990 by 2030
4	India	2,597 (6.8%)	+1.6%	1.9	No commitment to peak
5	Russia	1,793 (4.7%)	–0.8%	12.5	None listed on UNFCCC site
6	Japan	1,154 (3.0%)	–2.1%	9.1	–25% of 2005 by 2030
	Subtotal	25,490 (67.0%)			
7	Iran	702 (1.8%)	+3.4%	8.5	None listed on UNFCCC site
8	S. Korea	652 (1.7%)	–3.2%	12.7	–37% from BAU by 2030
9	Indonesia	626 (1.6%)	+8.0%	2.3	–29% to –41% BAU by 2030
10	S. Arabia	615 (1.6%)	+1.5%	18.0	No commitment to peak
11	Canada	585 (1.5%)	–1.4%	15.7	–30% of 2005 by 2030
12	South Africa	495 (1.3%)	+1.5%	8.5	398 to 614 Gt by 2025 or 2030
13	Mexico	485 (1.3%)	–1.6%	3.7	–50% of 2005 by 2030
14	Brazil	478 (1.3%)	–1.3%	2.25	–43% of 2000 by 2050
15	Australia	433 (1.1%)	+4.2%	17.3	–26% to –28% of 2005 by 2030
16	Turkey	416 (1.1%)	–1.5%	5.0	None listed on UNFCCC site
	Subtotal	5,487 (14.5%)			
	Total – 16	30,977 (81.5%)			
	A&S[3]	1,358 (3.6%)	+3.0%		
	All others	5,682 (14.9%)			
	World	38,017	+0.9%	4.9	

Notes to Table 6.2

1 The numbers and ranking in Columns A to E are taken from European Commission, Joint Research Centre, Emissions Database for Global Atmospheric Research [EDGAR], 2020 Report (EC 2020), table 1 and elsewhere. These numbers are fossil-fuel (EFF) emissions only. Figures for 2020 are anomalous because of the COVID pandemic and therefore are not used here.

   Although the EDGAR numbers for individual nations or regions represent fossil-fuel (EFF) emissions only, they may be regarded as a good measure for the discussion here, since among all anthropogenic GHG emissions sources, those from fossil-fuel use are, generally speaking, the most amenable to regulatory and policy direction at the national level.

2 Column F: UNFCCC-NDC. BAU = Business as Usual scenario.

3 International Aviation (730) and International Shipping (628).

GRAPPLING WITH 1.5°C WARMING

The radical resetting for the climate-forcing timelines is, of course, directly relevant to the treaty-dependent provisions that will be required to deliver emissions-reductions results on that basis. How difficult will it be to avoid this outcome? To assess this situation, the information assembled in table 6.2 must be considered carefully.

The long-term emissions trend lines (1990–2019) for the six largest CO_2 emitters, the countries or regions where annual emissions now exceed 1000 Mt (1 Gt) annually, are: China, +460 per cent; India, +433 per cent; Japan, +0.5 per cent; USA, +1 per cent; EU-28, –25 per cent; Russia, –25 per cent; World, +67.6 per cent (EDGAR 2020). The figure for Russia reflects in large part the effects of the collapse of the USSR in 1991. As indicated in table 6.3 below, the most likely future scenarios for the period 2018 to 2030 are that three (the United States, the EU-27+1, and Japan) will show decreases and the other three (China, India, and Russia) will show increases. Large increases for China and India, taken together, are likely to exceed the total of the decreases of any others by a wide margin, resulting in a significant net gain in global CO_2 emissions as of 2030 for the top six emitters collectively. However, only when one seeks to estimate the emissions projections for the countries among the ten largest second-tier emitters does the very real problem about the decade leading up to 2030 come into sharp focus.

Table 6.3 | Breakdown for 2019 global GHG emissions

Sources	Total	%	Per capita
G20 nations (22)	40.3 Gt	77.0	8.3 (tonnes/CO2 eq/ person)
"Other Large Emitters" (10)[1]	4.0	7.5	5.8
Rest of world (187)	6.7	13.0	2.0
International transportation	1.4	2.5	
Global total	52.4 Gt		

Notes

1 Iran, Egypt, Kazakhstan, Thailand, Vietnam, Malaysia, Nigeria, Taiwan, Ukraine, United Arab Emirates. China, the largest emitter, accounts for almost 27 per cent of the global total, but on a per-capita basis is only half of that for the leader, the United States.

These problems are only compounded when we look at emissions as tabulated in the other key measure, namely, all greenhouse gases. The two different measures are shown for the principal emitting nations in tables A.1 and B.1 in Olivier and Peters (2020), *Trends in Global CO_2 and Total Greenhouse Gas Emissions*. In table A.1, CO_2 only, the global total is 38.0 Gt; B.1, for all gases, $MtCO_2e$ is 52.4 Gt. Olivier and Peters then add a 5.0 Gt estimate for LULUCF for a grand total of yearly GHG emissions of 57.4 Gt for 2019. One should bear in mind the point made earlier: namely, that from a total-impact perspective related to human-caused global-warming pressures, CO_2 accounts for only two-thirds of the principal factors. This results in another way of indicating dramatically the current inequalities among nations making up global emissions. The material in table 6.3 is from tables B.1 and B.5 in Olivier and Peters, *Trends in Global CO_2 and Total Greenhouse Gas Emissions* (2020), which gives national figures for all GHG emissions ($MtCO_2e$/year).

Table 6.2 above had indicated that all the 151 nations ranked below the top sixteen emitters, and which are signatories to the Paris Agreement, were collectively responsible for only 15 per cent of global fossil-fuel

Table 6.4 | Top-sixteen estimated projections for GHG emissions (Gt) in 2030

A Rank 2030 (est.)	B Country	C GHGS 2019 MT (O & P)	D Est. 2030 GHGS	E Last NDC	F INDC Paris Pledge + Newer Commitments MtCO₂e	G Projected change 2019 to 2030
1	China	14,000	16,000	2015	Peaking "before 2030"	+2,000
2	USA	6,600	5,000	2021	–52% 1990 (7,100): Est. –27%	–1,600
3	India	3,700	4,700	2015	Reduce emissions intensity	+1,000
4	EU27+UK	4,300	3,500	2020	"at least" –55%/1990 (5,700)	–800
5	Russia	2,500	2,500	2020	CAT: No reductions	-0-
6	Indonesia	1,100	2,000	2021	NDC 1684–2034 MTCO₂E	+900
	Subtotal	32,200	33,700			+1,500
	% change					+4.6%
7	Brazil	1,200	1,800	2020	Estimated +50%	+600
8	Iran	950	1,425	None	No reduction pledge (est. +50%)	+475
9	Japan	1,400	1,000	2020	–25% 2005 (1,277)	–400
10	S. Arabia	700	1,050	2015	No reduction pledge (est. +50%)	+350
11	Mexico	800	1,000	2021	Estimated +25%	+200
12	Turkey	600	750	2015	No reduction pledge (est. +25%)	+150
13	S. Africa	600	750	2015	No reduction pledge (est. +25%)	+150
14	Australia	800	500	2020	–26 to –28% 2005: Est. –20%	–300
15	S. Korea	700	700	2020	–24.4% 2017	0-
16	Canada	800	700	2021	40 TO 45% below 2005: Est. –25%	–100
	Subtotal	8,550	9,675			+1,125
	% change					+13%
	Total	40,750	43,325		Top 16	+2,625
	% change				Top 16	+6.5%

See appendix 1 for the sources and calculations for Table 6.4

emissions of 38 Gt of carbon dioxide gas as of 2019. Unlike table 6.2,' table 6.3 uses GHG emissions for the expected top sixteen countries and projects these forward to 2030, which is a key target date for the Paris Agreement.

Table 6.4 suggests that, in terms of a collective attempt to reach a global peaking of GHG emissions by 2030, much depends on what happens in China and India – and, of the two countries, the case of India carries very large uncertainties. As noted earlier, in terms of cumulative emissions, India was responsible as of 2017 for only 3 per cent of global historical totals. In terms of per-capita emissions, the figures are still more telling yet: India has the lowest emissions in this category of any nation in the top sixteen – and merely a small fraction of those for the highest emitters. Thus, in terms of the welfare of its population, in its INDC (initial or intended nationally determined contribution) under the Paris Agreement, India had good reason to forecast its hope for strong GDP growth for the period from 2015 to 2030. A GDP growth of any magnitude requires more energy; under existing circumstances, the country would have little option but to rely for this on its large coal deposits. In this context, the 2021 document "Boom and Bust: Tracking the Global Coal Plant Pipeline," from the *Global Energy Monitor,* reports that new coal-fired energy plants are still being constructed, notably in China, and that hundreds of new coal mines are under construction around the world, with hundreds more being proposed. The COP26 meetings in Glasgow in November 2021 focused heavily on targets for phasing out the use of coal in particular and fossils fuels generally, but analyses have shown how difficult attaining this objective will be (Vinichenko et al. 2021). In fact, the International Energy Agency's *Coal 2021* report predicts that coal use for energy will reach record-high levels in both 2021 and 2022, based on strong demand in both China and India (IEA 2021d).

Referring again to table 6.4, only two countries/regions in the first set (USA, EU) and three in the second (South Korea, Canada, Australia) are fully developed modern economies whose NDC pledges show substantial announced future reductions. The remaining eleven are expected to have some substantial future increases in GHG emissions based on the pressures of economic development. In addition, analysis by Climate Action Tracker has identified a few countries standing not far below the top-sixteen-tier emitters, such as Argentina, Kazakhstan, and the United Arab Emirates, where there are also projections for significant increases by 2030. For most of the 150-plus countries that are currently at the bottom of the emissions list, the same pressures will affect both energy use and GHG emissions. *In summary, then, out of the total number of political entities (194) that are signers of the Paris Agreement, only five are guaranteed to have emissions reductions between 2019 and 2030.* All the twenty-four others named above, which may be labelled in relative terms as "major emitters," including the two that will be by far the largest by 2030 (China and India), are very likely to post increases in term of both CO_2 and GHG emissions for the same period.

The bottom line is that the world, taken as a whole, will almost certainly not be able to achieve a peaking of global GHG emissions by year 2030, a prediction that is supported by the recent work of the International Energy Agency referred to earlier. This prediction is now also supported by the latest release of information from the UNFCCC Secretariat, in a report dated 17 September 2021 and entitled "Nationally-Determined Contributions under the Paris Agreement: Synthesis Report by the Secretariat" (the report includes a nice graphical representation of its main point). The report includes "information from 86 updated or new NDCs submitted by 113 Parties. The new or updated NDCs cover about 59% of Parties to the Paris Agreement and account for about 49% of global GHG emissions." The analysis of the updated set of NDCs concludes that global GHG emissions are expected to be 55.1 (51.7 to 58.4) Gt CO_2eq in 2030, which is "59.3 per cent higher than in 1990, 16.3 per cent higher than in 2010 and 5.0 per cent higher than 2019." This is very close to my estimate for the top six emitters in table 6.4, and since they account for about two-thirds of global GHG emissions, this subtotal can be taken as a proxy for the world as a whole.

AVOIDING 1.5°C AND 2°C WARMING

A number of key considerations relating to the achievability of global warming targets are of vital importance. Fully half of all GHG emissions for the entire period from 1750 to 2011 occurred within the last forty years of that epoch, highlighting the most recent era as the apparent source of the world's current climate-change problems. However, as noted earlier, it is not only the recent GHG emissions that are driving climate change, but the total accumulation over more than a century. This raises the first of at least three fundamental equity issues in climate-change action, for it signals again the importance of the distinction between historic cumulative emissions and current ones. In 1990, the year often used in international negotiations as a baseline for emissions-reductions targets, the historical shares of emissions were as follows: the United States

(31 per cent), the EU-28 (30 per cent), and Japan (7 per cent), for a total of 68 per cent ("Cumulative CO_2 Emissions by World Region," Ritchie and Roser 2020, interactive graphic; see also Popovich and Plumer 2021). At that time, China's share stood at 5.36 per cent and India's at a mere 1.52 per cent.

It would therefore appear that the burden of any corrective actions applied specifically to the relative weighting on historic emissions would fall squarely on just those three nations and regions where significant emissions-reduction commitments in the NDCs have already been registered and expected: the United States, the EU-28 (following Brexit, EU-27+1), and Japan. However, it is very difficult to foresee how any further major adjustments could be made to these three NDCs in order to address this equity issue. (In the references section, see the article by Hof et al. 2017 on abatement costs for enhanced NDCs.) There are, of course, no enforcement mechanisms in the Paris Agreement, so any enhanced pledges would have to be voluntary.

The second equity issue is related to the fact that in the earlier stages of modernization all national economies have a higher level of energy intensity and GHG emissions relative to GDP than they do when they later develop greater energy efficiency. This is why making progress in this area is a part of some of the developing nations' initial NDCs under the Paris Agreement. Historical fairness demands that this be recognized as a legitimate contribution on their part to mitigation efforts. This is also why assistance from developed to developing nations has been stipulated in all stages in the UNFCCC processes (including the clean-development mechanisms, joint implementation, technology transfer, financial flows, and capacity building). But this unavoidable economic disadvantage can only be overcome slowly and doing so is expensive; consequently, this area is unlikely to be the source of any significant new achievements in emissions reductions.

The third equity issue also involves current emissions and is something that has always been explicitly recognized in agreements under the UNFCCC; namely, the marked differences around the world in per-capita GHG emissions. As shown in table 6.2, per-capita emissions among just the top six emitters in 2019 range from a low of about 2 tonnes (India) to a high of 15.5 tonnes (USA). In terms of historic emissions, per-capita CO_2 emissions in the baseline year 1990 were: United States, 20.14; Russian Federation, 16.12; EU-28, 9.34; Japan, 9.23; China, 2.06; and India, 0.70 (all units in tonnes). The issue of fairness is especially acute in the case of the continent of Africa, which has some of the lowest emissions (except for South Africa) and poorest peoples in the world, and faces some of the harshest impacts of global warming. There are no obvious or easy solutions to the accumulated inequities represented by large differences in per-capita GHG emissions that are relevant to the campaign against climate change.

These three inequities are to a large extent intractable. Looking forward to 2030, the parties' NDCs will be caught tightly in a vise, with redistributional pressures from equity issues on the one side, and, on the other, the possibility that there will be relatively few resources available from developed nations to respond to them.

There is one other relevant matter here, namely, the accuracy of the standard set of estimates on GHG emissions around the world. A strong argument has been made to the effect that there is serious under-reporting in this matter. In fact, global GHG emissions may be under-reported by something like 8-13 Gt CO_2e, up to 23 per cent more than now estimated (Mooney et al. 2021). Not surprisingly, most of the under-reporting is thought to occur among still-developing nations. If this allegation turns out to be further corroborated to any major extent, it will represent yet another complicating factor in the struggle to reduce emissions.

What this means is that future increases in GHG emissions from China, India, and other developing nations must be expected and cannot be amended in the short term, and that they are likely to outweigh substantially any new reductions made by developed nations, resulting on a net basis in a still-rising level of global emissions by 2030. The United Nations' stated objective of reducing emissions by 50 per cent from 2018 levels by 2030 is a pipe dream.

Given the NDCs submitted by these Parties and currently in place, passing the 1.5°C warming threshold by 2030, possibly by a wide margin, is unavoidable. Moreover, exceeding the 2°C warming threshold appears to be inevitable or at least very likely over the coming decades. Climate scientists have shown that the INDCs of the European Union, the United States, and China are consistent with a temperature increase of 3°C. The risk of overshooting 1.5°C puts the world on a trajectory to reach +2°C perhaps sooner than 2050. How serious would it be to pass the point at which a 2°C global temperature increase above pre-industrial levels occurs? Might a 2°C of global warming be the level at which humanity would be set on an unavoidable course to a catastrophic future? Here it is useful to repeat the quotation given earlier from a scientific paper published in 2018 (Steffen et al.):

We explore the risk that self-reinforcing feedbacks could push the Earth System toward a planetary threshold that, if crossed, could prevent stabilization of the climate at intermediate temperature rises and cause continued warming on a "Hothouse Earth" pathway *even as human emissions are reduced* [emphasis added]. Crossing the threshold would lead to a much higher global average temperature than any interglacial in the past 1.2 million years and to sea levels significantly higher than at any time in the Holocene.

According to these scientists, passing the threshold of a 2°C global average temperature increase is likely to set in motion what they call "tipping cascades," which are positive biogeophysical feedback loops – such as permafrost thawing, loss of sea ice, and release of frozen methane from oceans – that accentuate the trends in rising temperatures already occurring. Potential catastrophic effects following +2°C include sea-level rise of two to three metres or more by 2100, severe reductions in food output, extensive dieback of both boreal and tropical forests, and extreme heat. But the even more serious problem is that, once at +2°C, the climate system may become locked into the "Hothouse Earth" pathway, causing more temperature increases that will be irreversible, the effects from which will persist for millennia thereafter.

PATHS FORWARD

Pledges for specified reductions targets in the first commitment period of the Kyoto Protocol were made in 1997 only by a large group of countries in Europe along with Japan, the United States, and Canada. But the United States never ratified the Protocol and Canada first agreed and then withdrew, essentially leaving only Europe and Japan, which together accounted for less than 15 per cent of total global emissions. For Kyoto's second commitment period (2013 to 2020), only the European Union, plus Iceland, representing less than 10 per cent of global emissions, made a specific reduction pledge, that is, 20 per cent below their 1990 levels. Spanning the course of some seventy international meetings held between 1989 and the present, apart from the EU-28, Japan, and Iceland, *none of the other countries in the world, which taken together represent about 85 per cent of global emissions, has ever to date made and then held to a specific numerical pledge for reduced emissions.*

Three decades have elapsed so far in the overall UNFCCC process (1992 to 2022). Another timeline that may be more relevant is the twenty-five years from 1990 to 2015, when the Paris Agreement came into being, because, at the Paris meeting, parties for all practical purposes gave up on attempting to set binding targets for national emissions reductions, which had been the focus of the earlier Kyoto Protocol. Jackson et al. (2018) sum up as follows: "A quarter century after the United Nations Framework Convention on Climate Change, we remain far from its signature goal to 'stabilize greenhouse gas concentrations in the atmosphere at a level that would prevent dangerous anthropogenic interference with the climate system.'"

In the consensus views of climate scientists, as summarized above, there is already as of 2022 a fair amount of confidence in three key propositions:

1   Without more decisive and timely action, passing the +2°C warming level may be unavoidable.
2   There is a serious risk that passing the +2°C threshold may lock in irreversible further warming, regardless of emissions reductions undertaken thereafter.
3   Warming at +2°C and above may result in catastrophic outcomes.

As of now, the Sword of Damocles – representing the possibility that the world may not avoid arriving at, or even overshooting, the 2°C warming scenario – hangs over all the participants. When all is said and done, it is entirely possible that both Kyoto's partial top-down approach and Paris's inclusive bottom-up one will ultimately fail.

The first full "stock-taking" under the Paris Agreement, that is, the assessment of the progress made by nations in meeting the terms of their Nationally Determined Contributions (NDCs), is set for 2023, and the second for 2028. Probably not before 2028 will it be possible to make realistic assessments on two points: first, whether parties (especially large emitters) are on track toward fulfilling the terms of their NDCs; and second, whether those accomplishments, taken together, are consistent with the level of emissions-reductions trajectories needed to avoid exceeding the 2°C warming threshold. But the record to date does not inspire confidence.

In addition to India's situation, there is one other aspect of the scheduled stock-takings in 2023 and 2028 that overshadows all the rest, namely, what is happening in China. That country's NDC commits it to a peaking of GHG emissions "before 2030." This means that China's emissions, which are now 30 per cent of the global total, will continue rising until at least close to 2030; and, to the extent that many other nations proceed along the path toward reducing their emissions throughout the 2020s, China's share of the total will grow. Unless there is a radical change on China's part by 2023, at the first stock-taking, all prospects for the overall success of NDCs in meeting the goals of the Paris Agreement will dim appreciably.

Even though the United States returned to the agreement in early 2021, after a brief hiatus, it still resists making a legally mandated commitment to climate-change goals (as it has done since 1995, and unlike the European Union). This reduces the moral suasion that the United States can exert – and, it should be recalled, it is still (after China) the world's second-largest emitter. The great irony in all this is that, as a result of the impressive record of its economy in moving away from coal combustion and lowering energy intensity, the United States is very well positioned to meet its own announced first NDC target of 26 to 28 per cent below 2005 levels by 2025. However, there remains a high degree of political instability in the United States around the climate-change issue, and it is possible that its adherence to the Paris Agreement could be cancelled again in 2024. In that case, all bets on the survival of the entire international agreement would be off the table.

Should the 2023 stock-taking not result in a sense of overall strong progress in fulfilling the set of NDCs, and should China give no indication at that time of a willingness to change radically the nature of its commitment, five years later, in 2028, nations collectively will once again arrive at a crossroads. Parties will then squarely face the necessity of either revamping the Paris Agreement, almost certainly by strengthening all commitments (notably China's), or they will just hope for the best and soldier on as before. The conclusion that may fairly be drawn from the discussion so far is this: Unless the treaty process for climate change that has evolved to date is substantially strengthened with new national commitments, the current targets for GHG emissions control, as given by the science of climate forcing, cannot and will not be met.

In sum, the processes of treaty forcing and climate forcing have been fundamentally mismatched during the thirty-plus years from 1988 to 2022. For humanity, the terrible dilemma is that almost certainly there is no good substitute for seeking a solution to climate forcing through the existing mechanisms of international treaty law. But if the emissions peak does not occur before around mid-century, the situation we will likely be facing will be a drastic one. The 2017 US *Climate Science Special Report* states: "Without major reductions in these [GHG] emissions, the increase in annual average global temperatures relative to preindustrial times could reach 9°F (5°C) or more by the end of this century."

Despite the fact that climate change treaty-dependent timelines push right up against some potentially catastrophic impacts scenarios, there is the real chance that the world may never come to an agreement on an overall solution that involves binding, effective, verifiable, and enforceable sets of emissions-reductions targets for all major emitter nations and regions.

PROBLEMS WITH THE
TREATY-DEPENDENT TIMELINES

A brief historical review of important treaty negotiations during the second half of the twentieth century reveals two different, but equally important, problems: first, the length of time that can be required to achieve a satisfactory goal; and second, the need to determine whether a satisfactory goal is reachable at all. In three important other areas (chemical, bacteriological, and nuclear weaponry) the treaty negotiation process never did definitively resolve the most serious issues. This was because (a) with chemical weapons, important violators went unpunished; (b) with biological weapons, no verification was provided, and egregious violations went unpunished; and (c) with nuclear weapons, far too many stockpiles of such devices remain around the world, and the two nuclear superpowers continue to upgrade the danger and destructiveness of their arsenals.

Climate change too presents a difficult challenge – what has been called a "wicked problem" – for which no easy or clear solution presently exists. This is in part because delays, including century-long delays, are inherent in the very nature of this issue. The elements in the fight against climate change that cause delay include the long residence time of GHGs in the atmosphere; the lag between emissions and CO_2 concentrations in the atmosphere; the slow but relentless pace in the depletion of the allowable carbon budget, calculated from 1750 onwards; the postponing of the most obvious adverse impacts far into the future, when it will be impossible to avoid them; and, above all, the still-shrouded potential compound events or tipping points, when positive feedback loops might start to kick in, producing runaway escalation in further climate forcing. (On the last point see chapter 15, "Potential Surprises," in the US *Climate Science Special Report*, 2017.)

Ever since the lead-up period to the 1997 Kyoto Protocol negotiations, developed nations have known that whatever targets were specified for the first round of emissions reductions would be only a prelude to additional ones, and others also surely knew that they would be asked for their own commitments at some point. To some extent, this knowledge may have had an impact on their desire to postpone the inevitable.

The long delays in undertaking reductions programs, since Kyoto first proposed such targets in 1997, have provided much encouragement to those with a stubborn unwillingness to concede that there is even an issue that needs to be addressed at all. Such delays are the fertile breeding-ground for lazy skepticism about the prevailing scientific consensus, for cultivating the bizarre allegation that these scientists are motivated by the research money they generate, for dismissing all of the scientists' warnings as "fake news," for speculating on any cause of climate change other than human action, for wallowing in the abundant uncertainties that necessarily must accompany all risk assessments, for worrying whether the monetary costs of mitigation may outweigh the expected benefits, for thinking that, if we wait long enough, some easy technological fixes will reveal themselves, for believing that our descendants will find the solutions that have eluded us, for the conviction that others have caused the problem and should therefore bear the lion's share of responsibility for fixing it, for smugly pointing to earlier eons when the atmosphere was drenched in far more carbon than it is now, for dark musings about conspiracies, for a belief that whatever will happen is God's plan for us, and for any number of other idle farragoes.

To counterbalance them there is only the huge and complex output from thousands of qualified scientists, who hearken back to the nineteenth-century pioneers, Joseph Fourier, John Tyndall, and Gustav Arrhenius, and who readily concede that they cannot say with *absolute certainty* that rising anthropogenic GHG emissions are now the principal driver of climate forcing or that this forcing ultimately will have very bad outcomes for the current pattern of human settlements. Rather, they can only say that this process is "*extremely likely* to have been the dominant cause of the observed warming since the mid-20th century." The IPCC's Fifth Assessment Report (2014) states:

> Continued emission of greenhouse gases will cause further warming and long-lasting changes in all components of the climate system, increasing the likelihood of severe, pervasive and irreversible impacts for people and ecosystems. Limiting climate change would require substantial and sustained reductions in greenhouse gas emissions which, together with adaptation, can limit climate change risks ... Many aspects of climate change and associated impacts will

continue for centuries, even if anthropogenic emissions of greenhouse gases are stopped. The risks of abrupt or irreversible changes increase as the magnitude of the warming increases.

In fact, the majority of the key conclusions in the IPCC reports are now made with *high confidence* or *very high confidence*, based on the underlying analytical models, supporting data, and many years of seeking consensus judgments. The reader can find in figure 2 of the US 2017 *Climate Science Special Report* a useful summary of what are called the "confidence levels" for the various probabilities assigned to the conclusions and future projections of climate science.

But how likely is it that, at some point within the next twenty years or so, the climate scientists' judgments will finally win the day in the court of world public opinion? How likely is it that the majority of citizens will demand that their governments enact a verifiable and enforceable international treaty that is sufficiently robust to drive GHG emissions down in all of the world's major economies to the net-zero level by 2050? It would be nice to be able to say that we know the answer to *these* questions with any level of confidence at all.

7

Managing the Risks of Global Warming

About thirty years had elapsed earlier between the publication of the first prominent article on what would later be named climate forcing (Revelle and Suess 1957) and the highly publicized consensus among climate scientists that was achieved at the 1988 International Conference on the Changing Atmosphere held in Toronto. That conference in turn led, a mere four years later, directly to the United Nations Framework Convention on Climate Change, an international treaty signed in 1992 which came into force in March 1994 and eventually counted 197 signatories, amounting to all the member states of the United Nations. The follow-up international meetings designed to implement its provisions have lasted about twenty-five years: COP1 was held in Berlin in early 1995 and COP26 was held in Glasgow, Scotland, in November 2021.

The COP meetings have kept alive the promise that, eventually, an effective solution to address appropriately the risk of global climate change will be found. Across the preceding decades, the knowledge base underlying the broad scientific consensus on climate forcing has gone from strength to strength. At the same time, the solutions to climate-change risk laid down in the Paris Agreement have so far been failing to keep pace with new findings in the scientific literature about the pace of global emissions reductions needed to avoid dangerous anthropogenic interference with the climate system.

The mismatch between the treaty and the science was explicitly acknowledged in proceedings under the Paris Agreement called the Talanoa Dialogue, serving as the interim stock-taking of progress under the agreement: "Reports submitted to the 2018 Talanoa Dialogue indicate that the aggregate efforts from existing NDCs fall well short of achieving the long-term goal of the Paris Agreement." The solution for the deficiency was given in the "Talanoa Call for Action," a joint statement by the Presidents of COP23 and COP24:

> According to the science, global emissions continue to rise. This leaves a significant gap in the effort needed to limit global warming to well below 2 degrees Celsius above pre-industrial levels and to pursue efforts to limit the temperature increase to 1.5 degrees Celsius. The IPCC Special Report on 1.5 degrees highlights, among other things, the benefits of holding warming to below 1.5 degrees. It also concludes that to keep global warming within 1.5 degrees, global emissions need to be halved by 2030.

But the 2018 IPCC Special Report demonstrated that the existing set of NDCs will not come anywhere close to halving current emissions levels by 2030. One should not fail to note the irony that the need for a 50 per cent reduction from then-current levels was first mooted thirty years earlier at the 1988 Toronto Conference.

In the modern era, humanity has faced certain types of catastrophic risks, such as the Second War and the later threat of massive destruction and radioactive contamination from large-scale nuclear warfare. But as I have argued in earlier chapters, the world has never faced a risk such as climate change, for at least two reasons. First, climate change very gradually "loads" into the climate system the possibility of a number of truly dire impacts, but defers their appearance into some distant future – and impacts that are not now evident therefore do not seem to be of immediate relevance. Moreover, the magnitude of these projected impacts can be given only in the form of probabilistic estimates, so that it is possible to believe that they might never come to pass.

Second, this loading of likely future impacts is such that, given sufficient additional human-caused GHG emissions, the world may unwittingly cross a kind of invisible threshold, beyond which no subsequent mitigation efforts can prevent those catastrophic impacts from occurring. The science of climate forcing has squeezed mercilessly the timeframe within which the nations of the world can take effective action on climate change: As shown earlier, within just a few years the deadline at which the world should arrive at net-zero emissions was rolled back from 2100 to 2050. This has resulted in a steadily collapsing interval between achieving an effective agreement among nations on

GHG emissions reductions, on the one hand, and, on the other, the closing of the window of opportunity for avoiding dangerous interference with the climate system.

Some of the most recent climate-science journal articles present the consequences of this collapsing interval in the starkest terms. Lewis et al. (2019) remark that "none of the major emitters has made commitments [under the Paris Agreement] that are aligned with limiting warming to 2°C" and that current climate pledges "are estimated to result in a median global warming range of 2.6–3.1°C above pre-industrial levels." Jiang et al. (2018) state: "A mitigation pathway consistent with the Paris Agreement's 'well below 2°C' target requires halving gross CO_2 emissions every decade from ~40 $GtCO_2$ equivalent in 2020 to ~5 $GtCO_2$ in 2050 (the 'carbon law')." *This is functionally equivalent to the proposition that the Paris 2015 goals have little or no chance of being met without major enhancements of emissions-reductions pledges.* And time to remedy the situation is perilously short. That this type of mitigation pathway seems at present to be so utterly unrealistic, in the context of all the internationally negotiated steps taken in the past thirty years, is perhaps the best indicator of the ongoing dilemma that the world now faces with respect to climate change.

The major dilemma can be summed up simply in terms of one fact: as of 2019, 151 of the countries that have signed the Paris Agreement represent about 46 per cent of the earth's human population of 7.6 billion, but account for only about 15 per cent of its GHG emissions. Many of the 151 are still poor and underdeveloped, and there is little doubt that they will need to grow their economies and their energy usage in the future – importantly, as will one of the top-six emitters, India. Then there are the expected future emissions increases for China. How can this dilemma be overcome?

Some authors answer this question by suggesting that the world should abandon the attempt to create an effective international treaty altogether. A 2014 journal article by Cinnamon Carlarne discusses the following thesis: "The problem, simply put, is that the existing global paradigm, which aligns and consigns climate change within the boundaries of international environmental law, is flawed. This article challenges this paradigm and argues that climate change is an issue of such scale and complexity that it defies resolution through the constrained channels of an international environmental treaty." Some support for this standpoint is offered in an article by McEvoy and Cherry (2016):

If parties treat climate change as a collective-action problem that leads to free-riding, then the lack of incentives to cooperate along with a weak enforcement mechanism warrant a pessimistic outlook. But if parties are willing to act unilaterally, the menu of promising climate agreement architectures may not require mechanisms that prevent free-riding. The need for mutual coercion may be overstated, and by discounting non-pecuniary motives, the benefits of individual action may be undersold.

And yet, can the free-rider problem really be so summarily dismissed? And under what realistic scenarios do we imagine that what Carlarne has called elsewhere "less hesitant mitigation efforts at the sub-global level" can respond adequately to the enormous challenge represented by the need to achieve net-zero global emissions by 2050?

As discussed further below, the right approach is to build new multi-stakeholder initiatives around the supportive framework of the Paris Agreement. An important international actor, the Bank for International Settlements (BIS), has reinforced this theme in its discussion of "green swan risks," a name for climate-related risks, defined as "potentially extremely financially disruptive events that could be behind the next systemic financial crisis." The BIS states: "This complex collective action problem requires coordinating actions among many players including governments, the private sector, civil society and the international community." One major unknown in terms of private-sector players has to do with commitments to the 2050 net-zero emissions goal which might be made (and delivered on) by the largest corporate enterprises and investment-management firms. In early 2021, BlackRock Inc., the largest such firm in the world, advised the companies in which it held a stake: "We expect companies to articulate how they are aligned to a scenario in which global warming is limited to well below 2°C, consistent with a global aspiration to reach net zero greenhouse gas (GHG) emissions by 2050." Like the aspirations of governments, these expectations are promises of future actions which may or may not materialize.

In the remainder of this chapter, I will discuss the major options available to all nations, including Canada, for tackling climate change in the time remaining before 2050. The topics to be discussed are: (1) the "deep decarbonization" strategy; (2) geoengineering of the climate; and (3) technologies and policy options for GHG mitigation under the broad category of "carbon management" (carbon pricing, carbon markets, carbon sequestration, and carbon utilization).

THE DEEP DECARBONIZATION STRATEGY

Decarbonization is, quite simply, the gradual reduction in the use of fossil-fuel energy sources in a national economy. A more precise way of expressing this idea is to refer to declines in the economy's *emissions intensity* or *carbon intensity*, that is, the ratio between carbon-dioxide emissions and Gross Domestic Product. "Deep decarbonization" means removing all sources of such emissions, an objective that is consistent with the new target of net-zero emissions by 2050. The Deep Decarbonization Pathways Project, based in Paris, is a large consortium of nations devoted to the task of describing how to reach this objective. The three most important pathways are: efficiency and energy conservation; decarbonization of fuels and electricity; and a switch to low-carbon (ultimately zero-carbon) energy.

Efficiency means lowering emissions intensity through technical improvements in a multitude of sectors, including building design, urban planning, freight and passenger transportation, and construction materials; conservation requires extending the useful life of products and much-enhanced reuse and recycling. The fuels and electricity sector changes encompass moving to renewable energies, nuclear energy, and also fossil-fuel energy that is coupled with carbon capture and storage. Zero-carbon energy technologies include decarbonized electricity, biofuels, hydrogen, and synthetic natural gas.

In industrial economic development over time, decreasing emissions intensity is a structural outcome of production efficiencies, reduction of waste, lower energy prices, and other factors. To the extent to which fossil-fuel sources are the primary driver of the process of production, their contribution per unit of GDP must steadily decline over time. However, they have been the cheapest and most widely available sources of the types of energy required for initiating the process of industrialization during the past 250 years; this was true for the advanced industrial economies long ago and is equally so for today's still-developing nations. Increasingly, this is where all the emissions growth in the world is concentrated as of now. Jiang et al. (2018) comment: "The greatest mitigation challenges lie in the developing world. Real progress toward the Paris Agreement goal awaits an effective commitment by leading countries to undertake breakthrough research and development of low-, zero-, or even negative-carbon-emissions energy technologies that can be deployed at scale in the developing world." But even assuming that this kind of breakthrough research is successful, how is it to be paid for, given the capital constraints in the poorer nations – including by far the largest, India, but also others such as Indonesia and Nigeria – which are most in need of economic development?

A targeted strategy is urgently required in order to deal with this dilemma. The ever-shortening timeline for heading off possibly catastrophic impacts means that the mature decarbonizing process now under way in the developed world must be greatly augmented by having those nations provide to the others, gratis, massive quantities of non-fossil-fuel energy technologies as soon as possible. Other efficiency technologies that promise to reduce emissions/carbon intensity per unit of GDP could also be considered. This assistance will have the additional benefit of at least partially offsetting the historical inequity represented by cumulative emissions, the amounts of which are still dominated by the contributions of the already-developed countries.

The world requires a precise focus on what can reasonably be expected to happen, in terms of emissions in a large group of countries with "emerging and developing economies," over the next two decades. The International Energy Agency (IEA 2021e) predicts that this group will account for most of the GHG emissions growth in the world in the coming twenty years, perhaps amounting to as much as five gigatonnes of new emissions annually. I suggest that there is only one *realistic* scenario under which the world as a whole can realize the further emission reductions needed to avoid dangerous anthropogenic interference with the climate system. That is for decarbonization strategies to be widely adopted in all the still-developing economies, *using massive subsidies from the developed world over decades.*

One precisely targeted approach could be based on the pledge made by eighteen developed countries in September 2015, in the context of the negotiations leading up to the Paris Agreement: namely, to direct very significant funding to less-developed economies in order to avert emissions rises there. At Paris in 2015, a consortium of eighteen nations – Australia, Belgium, Canada, Denmark, Finland, France, Germany, Italy, Japan, Luxembourg, the Netherlands, New Zealand, Norway, Poland, Sweden, Switzerland, the United Kingdom, and the United States, plus the European Commission – affirmed "the goal of jointly mobilizing $100 billion dollars a year by 2020 from a wide variety of sources, public and private, bilateral and multilateral, including alternative sources of finance, to address the needs of developing countries, in the context of meaningful mitigation actions and transparency on implementation." (The $100 billion aid pledge was originally made in 2010: see WRI 2021a for a comprehensive breakdown and analysis by country.) An important document by Kevin Rennert et al. on "The Social Cost of Carbon," released in September 2021 by the Brookings Institution, is just the latest relevant document to argue forcefully that global climate-change targets simply cannot be met unless the efforts of poorer countries are adequately supported by wealthier ones.

However, as a 2020 report from Oxfam has shown, only much smaller sums have actually been provided to developing nations so far. The journal *Nature* noted in an editorial in early 2021 that United Nations Secretary-General António Guterres has publicly complained about how the developed countries have lagged in their pledge to provide US$100 billion a year in climate funding for poorer countries by 2020 (UN 2020). In fact, these sources have revealed the highly embarrassing fact that, in their reporting on climate finance, some developed nations are counting *loans* (up to 80 per cent of the total financing) toward their commitments! (Also, the Trump administration withdrew its support for the consortium pledge, although one might expect it to be reinstated by the Biden administration.) In late October 2021, Canada and Germany stated, on behalf of the secretary-general, that it is hoped the pledge first made in 2010 can be realized by the year 2023 (UN 2021). It still remains to be seen whether the mechanisms of assistance under the UNFCCC's climate-finance program will address at all adequately the decarbonizing issue in developing nations. Yet we should ask, *how* might it be made to work if the Group of 18 are soon going to get serious about their pledge?

A GLOBAL DECARBONIZATION BOND

A strategy of decarbonization is urgently required, and where that strategy is most needed is in developing countries, which are, however, constrained by a shortage of capital to carry it out. In order to deal with this problem, I propose that developed countries implement their promised $100 billion annual assistance to developing nations by issuing, by 2023, a $2.5 trillion "Global Decarbonization Bond" that will front-load the provision of appropriate technologies for this purpose. Devoting the promised assistance fund *entirely* to a single goal, namely, financing major technology acquisitions and other means for a broad program of decarbonization in developing nations, is a way of preventing the assistance from being frittered away in a miscellaneous array of small projects, which would all have to be administered separately.

The duty to provide assistance in this regard has been a regular feature of all proceedings under the United Nations Framework Convention on Climate Change since the treaty's beginnings in 1992. But both the tenor and the substance of this newer pledge reinforces the underlying strategy of the Paris Agreement: "polycentricity," that is, the engagement of a broad suite of actors beyond national governments. Any new proposal, such as the one described below, should take advantage of this strategy. How much money is needed and what kind of enlarged participation is possible?

According to Hannah Ritchie, major categories of carbon-abatement opportunities are as follows: energy efficiency, low-carbon energy supply, terrestrial carbon (agriculture and forestry), "technical measures," and behaviour change. If all these were to be taken up together, she has argued, a decrease of thirty-eight $GtCO_2eq$ in global GHG emissions could be realized, measured against a business-as-usual scenario that would otherwise result in emissions of seventy $GtCO_2eq$ at year 2030. Ritchie cites an estimate of the total global cost for utilizing all potential opportunities for abatement to be €200 to €350 billion per year by 2030, noting that this sum is "less than 1% of the forecasted global GDP in 2030." The author adds that the timing of the necessary investments in technologies required for such utilization will affect the variation in costs, with considerably larger sums required in early years, declining over the life of the investments. Ritchie concludes: "The upfront capital investment needed is €530 billion per year by 2020 and €810 billion by 2030. Although these figures may seem substantial, many estimates project that the economic costs of not taking action to avert climate change would greatly exceed investments in mitigation opportunities."

Clearly these numbers are very rough estimates and should be viewed only as indications of the scale of the problem. Also, the reason why Ritchie's analysis has much higher figures for the 2020s is that significant capital investments in decarbonized energy technologies are needed as early as possible in this decade in order to displace investments that will otherwise be made in new fossil-fuel energy plants – which have an average useful life of forty years. Ritchie correctly emphasizes the necessity of *front-loading* such investment so that its benefits in terms of decarbonization can begin to be realized before 2030.

The modest proposal advanced here is that the Group of 18 developed countries should use their collective pledge of $100 billion annually (indexed to the average inflation rate) to service a US$2.5 trillion Global Decarbonization Bond, backed collectively by the full faith and credit of those eighteen national governments. This means that, instead of averaging the support for each year, the entire commitment for, say, a ten-year period should be allocated at once, up front, for the simple reason that the sooner that decarbonized energy installations are up and running in the still-developing nations, the sooner will the benefits of reduced greenhouse-gas emissions be realized.

For such a Global Decarbonization Bond, let us assume a fixed 1 per cent interest rate and an adjusted principal, like the US Treasury Inflation-Protected Securities, which have a term of twenty-five years, with 2 per cent of the issue maturing each year. (The interest rate could be adjusted if required; these are nominal figures given only to initiate a discussion on the basic concept.) And here is a key aspect of the proposal: in the spirit of the Paris Agreement, this could make a whole host of private-sector parties and non-governmental organizations full partners in the fight against climate change in their role as investors. As a very secure investment, this bond would be attractive to sovereign wealth funds as well as to governments, private-sector businesses, and individuals. The Credit Suisse *Global Wealth Report 2020* estimated that aggregate private wealth as of the end of 2019 stood at US$399 trillion. Subnational entities –for example, a broad consortium of US states led by California and committed to decarbonization policies – could emerge as important actors in this enterprise. Conceivably, in the future, all or some of the twenty-five US states that are members of the US Climate Alliance might take some additional independent actions, complementing what is done by the federal government and supporting the objectives of the original Group of 18 pledge.

Having sold off the debt to investors, the Group of 18 countries would be left with only the annualized obligations for paying interest and retiring the bonds. Although in the past all the developed nations have separately provided various types of international-aid support to developing countries, their collective pledge on climate action was an unprecedented step, and it would not have been undertaken had the parties to it perceived that there were insuperable legal obstacles to its implementation. However, it falls short in failing to recognize the necessity for front-loading their investments in global mitigation actions. In this respect it is undoubtedly the 1948 Marshall Plan that provides the' most interesting precedent, for that plan allocated the quite astonishing investment of $100 billion (in 2018 USD) over four years in order to jump-start the economic recovery of Western Europe following the Second World War.

The largest share of the proceeds from the US$2.5 trillion Global Decarbonization Bond would be devoted to a program of building, shipping, installation, operator-training, and plant maintenance for a variety of non-fossil-fuel energy-generation systems, all produced by developed countries and provided gratis to developing ones. The recipient countries would be obliged to establish sinking funds to partially finance replacement facilities at the end of the useful life of the donated ones, and the more advanced among them might also be asked to provide some matching funds from their own resources.

As to the types of energy systems to be supplied, one might provisionally suggest a mix of wind-turbine farms, solar-panel farms, and primarily – to generate sufficient baseload quantities of reliable electricity – small-scale nuclear power plants. These plants could be the small Generation IV sodium- or lead-cooled fast reactors, which operate with a high level of safety and do not need to be placed along seacoasts or lakefronts. As producers and distributors of the facilities for all these energy technologies, the Group of 18 would share the high-value employment opportunities as well as the secondary economic benefits derived from all phases of the supply process, in proportion to their share of fiscal responsibility for the bond issue.

Unsurprisingly, the Group of 18 does not now propose to fulfill its 2015 pledge of $100 billion annually before 2023. But if they were to do so, they would find that the idea to implement it through a Global Decarbonization Bond would bypass the single greatest *political risk* in the pledge itself: namely, that later governments could renege on it entirely, or for one or more years, or in response to other significant events, such as wars or new pandemics which might cause nations to redirect the funds. The global bond, on the other hand, would legally commit a full decade of funding up front (although it is financed annually) and is therefore functionally irreversible. There are two other advantages. One is the requirement that funds are offered for a single purpose only, namely decarbonization, rather than being spread over a multitude of projects. The other is that the bond takes the form of a delivery of actual energy-generating facilities, not cash or subsidies for countries to purchase them directly. If some are offended by this strategy, they will be free to decline to participate.

Thus, this funding would be targeted at a specific form of assistance and be securely locked in for a long period of time. On the other hand, annualized contributions, fluctuating in line with short-term political and economic considerations, would be very unlikely to create a dedicated path to decarbonization in the developing countries. Unless the full commitment is front-loaded, so that all or most of the energy-technology resources are actually installed and operating by 2030, peaking of global GHG emissions will be delayed indefinitely beyond that point. Issuing this debt and starting production of the facilities relatively soon, say by 2023, when the first stock-taking of progress under the Paris Agreement will occur, is arguably the best way to head off the failure to reach the global 2030 target that is otherwise expected.

Even just delaying the commencement of this plan until the second stock-taking, scheduled for 2028, may result in its being too late to make a difference. On the other hand, there may be a sufficient level of concern, by 2023, about missing the Paris Agreement targets that the group of donor nations might be willing to double the amount of their collective pledge, providing a \$5-trillion bond. If this seems to be a shocking level of expenditure, it is worth noting that developed countries will probably incur something like this level of new debt within a mere two years in the struggle against COVID-19.

CLIMATE ENGINEERING
(SOLAR RADIATION MANAGEMENT)

Climate engineering or geoengineering refers to techniques designed to change the process of climate forcing on Earth by means other than restricting human emissions of greenhouse gases. The most important techniques are various types of solar engineering or solar-radiation management by reflecting back or blocking the amount of sunlight that strikes the planet's surface – especially by the injection of fine sulfate aerosols into the stratosphere. In prior years, other technologies have been suggested for this purpose, such as placing orbiting mirrors in space to reflect sunlight; also, techniques other than solar engineering have been explored, such as iron fertilization of the oceans (designed to increase the oceans' capacity to absorb and retain carbon dioxide). Two other techniques, carbon sequestration and carbon-dioxide removal from the atmosphere, are sometimes included in the category of geoengineering, but here they are regarded as being fundamentally different in character and will be discussed separately.

More recently, debates about geoengineering have focused almost exclusively on solar-radiation management using aerosols. In this, human action would mimic nature: the ejection of massive amounts of sulfate compounds, primarily sulfur dioxide and hydrogen sulfide, into the atmosphere due to volcanic eruptions partially blocks sunlight and cools the temperature of the earth's surface. This was illustrated dramatically in 1991 with the eruption of Mount Pinatubo in the Philippines. It was the largest such aerosol effect in the entire twentieth century and cooled the atmosphere up to 1.3°F for three years. Massive eruptions of this kind in the past include Tambora in 1815 and Krakatoa in 1883, both in Indonesia, and Laki (Iceland) in 1783–84. Tambora in 1815 resulted in the "year without a summer" in Europe in 1816. But the fine sulfate aerosols are soon washed out of the stratosphere in the form of acid rain. Therefore, any deliberate human application of these compounds for the purpose of cooling the earth, in what is known as stratospheric aerosol geoengineering (SAG), would certainly have to be repeated every few years.

The technology of delivery is relatively simple – for example, using high-altitude aircraft or balloons to distribute the chemicals in the stratosphere – but the application itself is not. Achieving the desired effect through SAG would require decisions as to when to apply the chemicals (what season), at what latitude and height, in what aerosol size, and in what chemical mixture. This technology has been discussed for more than a decade, but some very large uncertainties persist around calibrating the specific effects it could be expected to have upon application in the stratosphere. These uncertainties involve the magnitude and spatial distribution of radiative forcing and various aspects associated with it (temperature, precipitation patterns, and so forth). In other words, what would be the actual correlations between the application decisions, such as aerosol size and season, and the observed effects at the surface of the earth? To be sure, scientists will try to model these correlations before carrying out the experiment, but only an actual trial would answer some of these questions. On the other hand, since we are familiar with sulfate aerosol injection from volcanic eruptions, it is unlikely that any unanticipated adverse impacts from the trials would do permanent damage to our environment.

Ever since the appearance in September 2009 of the influential report issued by the Royal Society of London, *Geoengineering the Climate: Science, Governance, and Uncertainty*, there has emerged a consensus that the most serious and difficult issues associated with this topic are not technical in nature, but rather ethical and political. The first and most obvious one is: is this a way of avoiding responsibility for controlling GHGs? Is this meant to be a *substitute* for actions pledged by nations under the Paris Agreement? One good reply has been that, if the world failed to meet the necessary targets for reducing GHGs, or to make those targets sufficiently rigorous to avoid dangerous interference with the climate system, then sulfate aerosol geoengineering could buy the nations some more time to finally get their act together on carbon-dioxide emissions reductions. Using SAG as purely an interim measure would at least help the world avoid the utter absurdity of simultaneously allowing emissions to increase while depositing more and more chemicals in the stratosphere. In such a situation, the world would be entering a vicious cycle with an unknowable outcome.

And yet other ethical and political issues are more intractable. A 2019 summary of these issues will be found in a report from Harvard University's Belfer Center, *Governance of the Deployment of Solar Geoengineering*, in which the key questions are given as follows:

1   *Who* ought to and/or will *specify criteria* for SG deployment, and
     who ought to and/or is likely to decide when the criteria
     are satisfied?
2   *What* will or should these criteria be?
3   *How* should/will decisions about deployment be made; what
     decision-making process should/will be utilized?
4   *What institutions*, either existing or new, are appropriate as
     decision-making venues? What will or should be the legal
     framework of such institutions?

Only a few of the key possibilities can be mentioned here. First and foremost, there is the risk that a powerful individual country might embark upon SAG initiatives on its own, despite objections from others. This is called a "free-driver problem" (where some can impose potential adverse effects on others who have not been consulted) as opposed to the better-known "free-rider problem" (where some benefit from the efforts of others without participating). Such single-country initiatives could give rise to "counter-engineering," that is, actions to nullify the efforts – including war, of course.

Other risks are distributive (some countries may experience more effects than others, or even effects that are critically adverse); intergenerational (the current generation may benefit at the expense of later ones); uncorrectable (SAG enterprises might entail unexpected adverse effects that cannot be remedied); ecological (SAG enterprises might entail unexpected adverse effects on ecosystems); procedural (a lack of adequate governance over SAG enterprises or the exclusion of some from decisions that affect them); injustice (the denial of equality of treatment to different populations); inadequately protective (when only the poorest and most vulnerable peoples are adversely affected); and others. These risks are widely acknowledged as factors that will have a significant bearing on whether SAG is ever deployed and, if so, under what conditions. Many research initiatives are now under way devoted to trying to determine how to manage these risks transparently and appropriately, including proposals for small-scale tests or laboratory studies to be undertaken prior to any consideration of full-scale deployment.

Under the heading of "governance," there are many discussions and research projects that are seeking ways to manage these risks. Considered as an alternative response to the difficult challenges in controlling climate-change risks, climate engineering or geoengineering might become an integral part of negotiations for a new, comprehensive climate treaty to replace the Paris Agreement or, alternatively, might become the subject of a separate international treaty dedicated solely to that issue. Organizations are now involved in discussing these suggestions with UNFCCC officials, IPCC, the United Nations Security Council, and other relevant bodies.

Given the uncertainties around climate-engineering technologies, and the magnitude of potential adverse consequences, adding this new dimension of risk to the already complex negotiations associated with the management of climate change within the framework of a single international treaty might prove to be very difficult. And yet it is hard to understand how climate engineering might be deployed acceptably and relatively safely except under the umbrella of such a treaty. One suspects that, if the need for solar-radiation management becomes acknowledged by national authorities around the world, they will attempt to regulate it through such an instrument. The risk of failing to do so, and of leaving deployment entirely in the hands of individual states, would seem to be clearly unacceptable.

CARBON MANAGEMENT
(CARBON PRICING AND REGULATION, CARBON SEQUESTRATION, CARBON REMOVAL)

1. Carbon Pricing and Regulation

Establishing a price for carbon is widely regarded as an important policy step in reducing greenhouse-gas emissions. Carbon pricing means imposing a cost for emissions of carbon dioxide on industry sectors or on some of the goods consumers purchase, such as fuel. For example, as of 1 January 2021, industries and citizens in all provinces of Canada were subject to the federal Greenhouse Gas Pollution Pricing Act, which sets a current minimum price of $20 per tonne for carbon, with incremental rises to $50 per tonne in 2022 and $170 per tonne in 2030. Provinces can set a higher price, and provinces can also impose a separate tax on a portion of the GHGs emitted by large industrial plants, the application of which is regulated by the federal government.

Consumers may be directly affected by what is called a "carbon tax" in the price they pay for fuels at gasoline stations. At the lower levels of the price for carbon this might be more accurately called a carbon levy than a tax, because families can receive rebates on their income taxes equal to or somewhat more than the increased amount they paid at the fuel pump. In this case the carbon levy would not be expected to have a significant impact on consumer choices, since the increased cost of fuel is offset by the annual rebate. Thus only the heaviest consumers of fuel might be motivated to change their choices and behaviour. When carbon is appropriately priced, governments and many analysts have argued, such pricing is the most effective and efficient policy instrument for securing reductions in greenhouse-gas emissions (ECCC 2018, 2020; Rivers and Wigle 2018; CICC 2020). Coyne (2021a) summarizes the case well and includes a very valuable chart entitled "The Cost of Not Pricing Carbon: Selected Non-Price Approaches to Reducing Greenhouse Gas Emissions."

At present two provinces, Quebec and Nova Scotia, use an alternative system called "cap-and-trade," which is designed to set a price for carbon emissions comparable to a carbon levy. In his book *The Citizen's Guide to Climate Success*, Mark Jaccard describes this policy as follows:

> Governments set a cap [limit] on emissions for all or some sectors of the economy, and auctions or freely allocates (called "grandfathering") tradable emissions permits (also called "allowances") that sum to the total emissions cap. In future years, the cap declines according to a schedule, meaning that the number of permits issued by government each year also declines.

Under this system, for example, an industrial plant that cannot meet its CO_2 limit in a particular year can purchase offsets at a "market" price from another company that has lower emissions than its own cap; the point is to set a limit for that industry sector as a whole. As the cap declines, due to government policy and regulation, the market price of allowances rises, making it more expensive for plants or sectors to fail to meet their limits, thus incentivizing compliance. A well-known example is California's Emissions Trading Program, administered by its Air Resources Board since 2013, which now covers 85 per cent of all GHG emissions in the state. The European Union currently operates the largest and oldest (since 2005) emissions trading system in the world.

The objectives underlying a carbon tax or levy can also be met using regulation, and if regulations are "flexible," rather than rigidly prescriptive, they can mimic the efficiencies of emissions trading. An example is a low-carbon fuel standard, which – as Jaccard explains in his *Citizen's Guide* – "requires that the average carbon intensity of energy sold for use in transportation decline over time." This imposes a declining carbon cap on individual fuel producers, but allows trading of permits between firms; in this way it acts like a cap-and-trade scheme. So, for example, higher-carbon oil fuel producers can purchase credits from electricity, biofuel, or hydrogen fuel suppliers. Other examples are the renewable portfolio standard for the electrical sector (requiring a prescribed amount of energy from renewable sources) and a vehicle-emissions standard.

2. Carbon Sequestration

Sequestering carbon means isolating it and storing it securely away from the atmosphere. The main strategy in this area is known as "carbon capture and storage" (CCS). Here CO_2 is stripped out of waste streams, such as the flue gas from coal-fired energy plants, by means of chemical absorption ("chemical scrubbing"), then liquefied under pressure, transported via pipeline, and sent deep underground. Other emitters of large volumes of carbon dioxide are natural gas fields, plants for coal gasification, pulp and paper, and nitrogen fertilizer, and other facilities. Carbon capture can be carried out either pre-combustion (gasification) or post-combustion.

The production of hydrogen from natural gas and its use as an energy source is a significant part of a zero-emission economy. So-called "blue hydrogen" is produced by stripping carbon out of natural gas and sequestering it. "Green hydrogen" is produced by electrolysis of water using renewable energy sources. The underground cavities in which the carbon is stored are geological formations that have been selected for their capacity to retain large volumes of liquefied CO_2 with minimal leakage over long periods of time. A fair number of capture-and-storage operations are already in place around the world, for example in

Australia, Algeria, Norway, the United States, and Canada. Research projects have assessed the risks associated with CCS and have concluded that they appear to be quite manageable (see IJRAM 2019 in the references). Forecasts have assigned a high degree of importance for CCS in overall strategies for reducing global carbon emissions, but the development of the requisite facilities so far has not been nearly as rapid as had been hoped (Martin-Roberts et al. 2021).

Finally, a process known as "Direct Air Capture," or carbon removal, seeks to sequester CO_2 from ambient air rather than from point sources – using chemical scrubbers – and then to store it permanently, as above. There are start-up firms developing this technology in Canada, the United States, and Europe. The scope of its eventual usage will depend, among other things, on the price of carbon. Carbon removal-and-sequestration is part of what is known as "negative emissions technologies," and they will be important, because they can be more efficient and cheaper alternatives to reducing emissions in some sectors (NAP 2019, NET 2020).

3. Carbon Utilization and Carbon Recycling

Removing and sequestering carbon and storing it underground means that we have no use for it and thus need to store it securely and indefinitely, ensuring that it does not escape into the atmosphere over very long periods. Carbon utilization, on the other hand, presupposes that those industries and their products that utilize this atom are indispensable for today's society. Carbon is the fourth most abundant atom in the universe by mass (after hydrogen, helium, and oxygen), and its truly remarkable properties in chemical bonding, forming a huge number of organic compounds under the ordinary pressure and temperature conditions on the surface of the earth, makes it the common element of all known life.

Carbon utilization (also known as carbontech), then, is based on the recognition that we must have the capacity to *recycle*, rather than confine underground, a great deal of the carbon already contained in the industrial products we use. When carbon-based molecules are incorporated into products, they are kept out of the atmosphere, where they would have a warming effect. (In other words, what we have to avoid is not carbon itself but rather further exploitation of "fossilized" carbon, that is, the vast coal and oil deposits sequestered earlier by nature deep beneath the earth's surface.) Targeted products include building materials, fuels, concrete, aggregates, and plastics, but the list of carbon-based products is practically endless. The basic idea is to "embed" carbon into materials directly and through their underlying constituents (polymers). The simplest example is cement: A Canadian company, CarbonCure, injects a stream of CO_2 gas supplied by other firms into the cement manufacturing process, where it actually improves the performance of the final product. Increasing the carbon content of products sequesters that carbon for their useful lifetimes.

Part of a large-scale process of this kind involves using a network of pipelines to move carbon around, from where it is being newly generated in industrial settings to where it can be embedded in new products. The second step is to maximize the amount of recycling of those materials once the products are worn out. The basic concept is to create, so far as possible, a closed loop for carbon-based materials. In other words, this is a way of adding "recarbonization" to the objective of decarbonization. It concedes that industrial society is and will remain highly dependent on the use of carbon-based materials. The new proposed practice is a radical departure from the prevailing one, in which carbon-intense products are made from hydrocarbons recently dug out of the earth, then vented out of smokestacks or discarded in landfills, whereupon new volumes of hydrocarbons are extracted, and so on.

8

Canada: Mitigation, Impacts, and Adaptation

The discussion so far has focused on Canada's commitments to other nations on the world stage as they pertain to controlling its GHG emissions. These commitments are twofold, both imposed under the Paris Agreement: our own reductions targets for 2030 and 2050, and our pledge to assist developing countries in their obligations on this score. Our record to date is not promising. In the three decades starting in 1990, we have missed every target we have set for reducing emissions. And yet seeking to assign blame for these failures would be pointless; neither Canadian politicians nor most citizens have tried hard enough in the past to honour our commitments for reducing our greenhouse-gas emissions. The critical moment for dealing with climate change has now arrived for both citizens and politicians. It is time for all of us to stand up and be counted.

In this chapter, we will turn our attention inward, to the aspects of the climate-change problem that are particular to our country. This has to do with the features of global warming that will directly impact the lives of Canadians in the coming decades. There are three aspects of dealing with these features. The first is *mitigation*, which involves the changes that must be made to our lifestyles and economic sectors as a result of our pledges to the international community to reduce our greenhouse-gas emissions from all sources by 2030 and to eliminate them completely by 2050. The second is *impacts*, that is, the expected effects that climate change will have on our land surfaces and coastlines as a result of increasing global average temperatures, especially because the warming will be much higher than average in the Arctic. The third is *adaptation*, which is the longer-range alterations in coastal habitats and economic sectors, especially agriculture and forestry, that will be required as the climate continues to warm.

A. MITIGATION

In the climate-change context, mitigation refers to a process of reducing human-caused emissions of greenhouse gases, or increasing the sinks where carbon is naturally captured – for example in forests – for the purpose or slowing or stopping global warming. Canada's original GHG emissions target for 2030 (–30 per cent/2005) was to cause the country to plateau at an annual rate of 511 megatonnes of carbon-dioxide-equivalent emissions. When the COVID pandemic struck, however, it had a marked effect on economic activity and energy production, and thus also GHG emissions, that lasted throughout all of 2020 and well into 2021 (EDGAR 2021); for example, Canada's emissions fell by 11 per cent in 2020, considerably higher than the global average of 7 per cent. But the global "rebound" was quick in coming (Friedlingstein et al. 2021, and Jackson et al. 2021). As Jackson et al. (2021) state:

Global fossil CO_2 emissions in 2020 decreased 5.4%, from 36.7 Gt CO_2 in 2019 to 34.8 Gt CO_2 in 2020, an unprecedented decline of ~1.9 Gt CO_2. We project that global fossil CO_2 emissions in 2021 will rebound 4.9% (4.1% to 5.7%) compared to 2020 to 36.4 Gt CO_2, returning nearly to 2019 emission levels of 36.7 Gt CO_2. Emissions in China are expected to be 7% higher in 2021 than in 2019 (reaching 11.1 Gt CO_2) and only slightly higher in India (a 3% increase in 2021 relative to 2019 and reaching 2.7 Gt CO_2). In contrast, projected 2021 emissions in the United States (5.1 Gt CO_2), European Union (2.8 Gt CO_2), and rest of the world (14.8 Gt CO_2, in aggregate) remain below 2019 levels.

UNEP's *Emissions Gap Report 2020* expressed hopes that, as the pandemic slowly receded and nations entered the economic rebound phase, the world might see a "green recovery," that is, an emphasis on less carbon-intensive practices in energy use. However, the report also conceded: "The COVID-19 crisis offers only a short-term reduction in global emissions and will not contribute to emissions reductions by 2030 unless countries pursue an economic recovery that incorporates strong decarbonization."

In fact, the World Meteorological Organization noted in October 2021 that by the end of 2020 global GHG emissions levels had begun rising again faster than the average rate for the past decade (WMO 2021); *rates of increase* for both methane and carbon dioxide in 2021 were the highest since measurements began thirty-nine and sixty-three years ago, respectively (NOAA 2022).

UNEP's *Emissions Gap Report 2021* contended that, on the basis of current NDCs, the world will fall far short of needed reductions in GHG emissions by 2030. In late March 2022, the release of IPCC's Working Group III report, *Mitigation of Climate Change*, gave more precise indications: (1) "current NDCs to 2030 collectively would barely reduce global emissions below present levels"; (2) "In the scenarios we assessed, limiting warming to around 1.5°C (2.7°F) requires global greenhouse gas emissions to peak before 2025 at the latest, and be reduced by 43% by 2030." In all likelihood, neither of these requirements can possibly be fulfilled.

Canada's experience in the next few years will reveal its own performance in this regard. For now, we are forced to speculate on what is likely to occur. Developed countries such as Canada are likely to put a high priority on getting back to the status quo ante, that is, economic performance as it was before the pandemic struck (which Canada had just about achieved by July 2021). One important reason for this expectation is that like other nations Canada increased both its federal and subnational accumulated public debt considerably, as both levels were forced to try to spend their way out of an economic collapse in order to keep both citizens and businesses afloat. The creditworthiness of both levels of government will suffer in the future as a result. An attempt to resume a strong economic growth path is the way in which Canada and others can hope to begin reducing their debt-to-GDP ratios once again. Thus the most reasonable expectation is that any effort to promote a "green recovery" in 2021 and 2022 will be on a modest scale at best, and that the existing technologies through which economic activity is transmuted into GHG emissions will largely be regenerated.

It is also likely that full and sustained post-pandemic economic recovery will not be in place until 2022. Therefore, the most probable short-term pathway for Canada's GHG emissions is that, as year 2023 opens, our emissions will have returned to something close to the 2019 levels, the last year for which there is as of now a certified total (*Canadian Energy Outlook 2021*, 203). If this turns out to be the case, Canada will have eight years left before 2030 in order to achieve its newly promised GHG reductions from 730Mt to 420Mt, a 42.5 per cent total drop and an average annual reduction of almost 39 Mt. Not to put too fine a point on it, this will be an onerous obligation, one requiring sustained dedication to that objective on the part of federal and provincial governments, the private sector, and individual citizens. Recall from chapter 5 that Canada's very first emissions-reduction pledge, made by Prime Minister Brian Mulroney in 1988, would have stipulated annual reductions of 17 Mt: The difference is the wages of procrastination.

One point is especially relevant. Canada has a long practice of alternating its federal governments between Liberal and Conservative regimes. Given this pattern, it is entirely possible that a Conservative government could replace the current Liberal one in the second half of this decade. During the twenty-first century to date, Conservative governments – with a strong representation from the oil-producing provinces – have been unsympathetic, to say the least, toward dedicated actions to reduce GHG emissions. In March 2021, at a national policy conference of the Conservative Party, a majority of delegates with strong support in the Western provinces voted *against* a resolution stating, "we recognize that climate change is real." Thus it is possible that, as 2030 approaches, with reductions still not yet on target, Canada will once again be in a position of not fulfilling its pledge. If this should happen in 2030, the situation will be much more serious than it had been in the past, and it will also mean that Canada will be nowhere close to net-zero by 2050.

Our commissioner of the environment and sustainable development released a report in November 2021, *Lessons Learned from Canada's Record on Climate Change*, with some harsh statements: "Canada has consistently failed to meet its emission reduction targets ... Past inaction on climate change has created the present crisis ... Canada's greenhouse gas emissions have increased since the Paris Agreement was signed, making it the worst performing of all G7 nations since the 2015 Conference of the Parties in Paris, France."

Canada's Emission-Reductions Strategy as of 2020

Every mitigation scenario for every country depends on an economic-sector analysis. There is an elaborate sector analysis for the world as a whole in the report by the World Resources Institute, *State of Climate Action 2021* (WRI 2021b); here we focus on Canada. Dramatic changes were made in the sector reductions scenarios for Canada in the 2021 edition of *Progress towards Canada's Greenhouse Gas Emissions Reduction Target* and its associated documents from 2020, *A Healthy Environment and a Healthy Economy* plus its *Modelling and Analysis Annex*, and

Table 8.1 | Expected emissions reductions in 2030 by sector (ECCC)

	Historical				Projected		Change 2005 to 2030
	2005	2010	2015	2018	2020	2030	
Oil and gas	158	159	191	193	177	138	–20
Electricity	119	96	81	64	38	11	–108
Transportation	161	168	172	186	155	151	–10
Heavy industry	87	75	79	78	65	61	–26
Buildings	86	82	86	92	90	65	–21
Agriculture	72	68	71	73	73	74	2
Waste and others	46	42	41	42	39	31	–15
LULUCF, NBS, agriculture	NA	11	-8	–13	–25	–27	–27
Total (incl. LULUCF)	730	702	712	716	612	503	–227

Canada's Greenhouse Gas and Air Pollutant Emissions Projections 2020. Table 8.1 is from the *Modelling and Analysis* document (for a good short exposition see Hughes 2021).

The most important initial observation is that the projected total for 2020 (637 Mt, excluding LULUCF) shows about a 13 per cent reduction from the 2019 level and that there is no acknowledgment in the table itself that this is all or almost entirely the effect of the pandemic, although this effect is mentioned elsewhere in the neighbouring text. The fact is that there will be no reliable accounting of GHGs for Canada and many other countries – one that cancels out all pandemic-related GHG reductions – until the 2022 figures are finalized sometime in 2023 or, if the pandemic persists, even one year later, in 2024.

The most dramatic general conclusion from the above table is that over half of Canada's emissions are attributable about equally to just two sectors: transportation and oil and gas. (One of the main common indicators for all countries is the transportation sector; WRI [2021b] estimates that this sector, globally, accounts for 17 per cent of all emissions.) Canada, as the country with the world's second-largest landmass and a relatively small population, is heavily reliant on all forms of transportation for people and goods; it also, of course, has a large oil and gas production industry. These specific sectors are prime candidates for changes under the nation's emissions-reductions targets. And yet the projected reductions by 2030 in these two sectors in table 8.1 are quite small (in the oil and gas sector, all the reductions are in the natural-gas subsector, and none are in the crude-oil subsector).

"Heavy Industry" is also a very important sector, accounting for no less than one-third of all GHG emissions worldwide, where new technologies are urgently needed (see J.H. Wesseling et al. 2017, J. Rissman et al. 2020, and M. Barecka et al. 2021). Cement manufacturing alone, for example, is responsible for 8 per cent of worldwide GHG emissions; producers, including some in Canada, are already introducing new technologies to reduce or eliminate those emissions, although significant challenges remain (Miller et al. 2021). Reducing emissions from steel production, which account for 7 per cent of the global total, also entails major challenges (Vogl et al. 2021).

The goal for the Electricity sector is a very aggressive one. For the period 2005 to 2030, this sector is responsible for almost half of all emissions reductions, and by 2030 the goal for the entire sector is forecast to be very close to zero GHG emissions. Since 82 per cent of electrical power in Canada already comes from non-emitting sources (hydroelectric, nuclear, and renewables), further progress in eliminating GHG emissions in this sector is a challenging objective. Strong policy direction at the federal level will be absolutely necessary, as Jaccard and Griffin argue in their 2021 report, *A Zero-Emission Canadian Electricity System by 2035*, because a full doubling (or more) of clean electricity supply in Canada will be needed in order to enable the transition from fossil-fuel use in the transportation, buildings, and industry sectors.

Finally, the Agriculture sector may do better than is forecast. A group called *Farmers for Climate Solutions* has offered to work with governments to make significant progress on emissions reductions in this sector, and there are opportunities for better sequestering of GHGs in Canada's vast grasslands as well. A major 2021 journal article by Drever

Table 8.2 | Partial provincial breakdown for Canada's emissions

Province	2019 (Mt)	2030	Change	% change
Quebec	84	79	–5	–5%
Ontario	163	162	–1	nil
Saskatchewan	75	66	–9	–12%
Alberta	276	250	–26	–9.5%
Total Canada	730	674	–56	–7.7%

et al., "Natural Climate Solutions for Canada," provides the first comprehensive and quantitative tally of potential carbon-emissions reductions from croplands, grasslands, wetlands, crop selection, manure, forest management, and tree planting. Three other recent studies of peatlands (Harris et al. 2021; see WCSC 2021) and stored soil carbon in Canada (Sothe et al. 2021, 2022) are relevant here. Sothe et al. (2022) comment that "Canadian soils and peatlands are assumed to store approximately 20% of the world's total soil carbon stock" as well as a full 25 per cent of the stock in the top one metre of soil (Sothe et al. 2021; see also Semeniuk 2021b), where the vulnerability of releases due to climate change is an important concern.

Living up to our professed commitments has always been, and will remain, largely a matter of political will. Our current federal MPs are today making promises for a decade hence, a point when many of them will be gone. It is far too early to declare victory in this battle, as some, such as the Canadian Institute for Climate Choices, have already done. And there remains one particularly significant hurdle, one that is eminently political: the distribution of expected emissions reductions among provinces and territories. As of 2019, for example, Alberta (38 per cent) and Ontario (22 per cent) combined represented 60 per cent of Canada's emissions of 730 Mt (*National Inventory Report, 1990–2019*, table ES-4). Now compare this with the projections by province for year 2030 in table 8.2 (*Canada's Greenhouse Gas and Air Pollutant Emissions Projections 2020*, section 1.3.5, table 9).

Thus, between 2019 and 2030, Alberta and Saskatchewan were expected to contribute 62.5 per cent of the nation's total emissions reductions for that period, whereas Ontario's total remains pretty much unchanged and Quebec has quite a bit less to do. To be sure, the numbers for Ontario reflect that province's earlier success in eliminating coal-fired electrical-generation plants. Yet the remaining differences among the major provinces in the distributional impacts in the energy, industry, and transportation sectors will have to be tackled through carefully designed government programs.

The expectations built into our climate plan up to 2020 were based on "64 strengthened and new federal policies, programs and investments to cut pollution." (They are listed in *A Healthy Environment and a Healthy Economy*.) And here is the decisive sentence: "The proposed actions outlined in this plan will – *once fully implemented* – enable Canada to exceed its current 2030 target" (italics added). (To be sure, the exceedance was trivial.) Thus, our hoped-for success in meeting even our original pre-2021 target for 2030 (–30 per cent/2005) was entirely dependent on a set of policies and measures not yet implemented. And now the target is far more onerous.

Sector Analysis of the Enhanced Targets as of 2021

As noted at the end of chapter 5, at a climate summit on 2021 Earth Day (22 April), the US president, Joe Biden, almost doubled that country's initial emissions-reduction target under the Paris Agreement, which earlier had been set at –27 per cent/2005 by 2025, raising it up to –52 per cent/2005 by 2030. Prime Minister Justin Trudeau responded by increasing Canada's 2030 target from –30 per cent of 2005 emissions levels to between –40 and –45 per cent. Meeting the original target would have put Canada's emissions in 2030 at 511 Mt; assuming that the midpoint of the new target is achieved (–42.5 per cent), 2030 emissions would be 420 Mt. Canada formally incorporated this new 2030 target into its enhanced "Nationally Determined Contribution" under the Paris Agreement, submitted to UNFCCC on 12 July 2021, stating (Canada, 2021 NDC): "Factoring in the Government of Canada's 2021 budget measures and additional actions, such as continued alignment with the United States, Canada's 2030 emissions would fall to 468 Mt (equivalent to at least 36% below 2005 levels)." This statement is highly ambiguous owing to its reference to unspecified "additional actions." As the October 2021 report by the Canadian organization Clean Prosperity emphasizes, everything depends on Canada's ability and willingness to actually implement the new promises that have been made, especially its reliance on carbon pricing (on this point see also Coyne 2021b).

Canada's sector-specific reductions targets for the newest and far-more-ambitious commitment (–40% to –45% of 2005 emissions levels by 2030) were set out in late March 2022 (ECCC 2022, Section 3.2). The projection, set at the lower end of the range (–40 per cent), yields a 2030 emissions total of 443 Mt. The seven projected sector reductions, calculated as a percentage drop between the years 2005 and 2030, which may be compared with table 8.1 above, are as follows: Oil and Gas (–31 per cent), Electricity (–88 per cent), Transportation (–11 per cent), Heavy Industry (–39 per cent), Buildings (–37 per cent), Agriculture (–1 per cent), Waste and Other (–49 per cent). These are aggressive targets for all but the Agriculture sector, and will require strong policy support and direction. For example, even though the reductions target for the Transportation sector is relatively modest, it still requires 60 per cent of all vehicle sales, and 35 per cent of all commercial truck sales, to be of electric vehicles by 2030.

An independent organization, the Trottier Energy Institute at Polytechnique Montréal, had earlier done a sector analysis in a report entitled *Canadian Energy Outlook 2021: Horizon 2060*. In general, the Trottier Institute report presents the most thorough set of economic sector scenarios for realistic emissions-reductions possibilities in Canada that has been prepared to date. The report focusses on actual policies now being implemented, rather than proposed or unspecified ones, and concludes provocatively (103) that these policies will reduce Canada's emissions by only 16 per cent below 2005 levels by 2030, far below even the original target of -30 per cent: Canada's target, it says, remains "out of reach with the measures announced so far." (In addition, there is another recent and shorter sector study, issued by the Royal Bank of Canada in October 2021, *The $2 Trillion Transition: Canada's Path to Net-Zero* [RBC 2021], which is presented in clear and nontechnical language.)

The Trottier Institute report contains two issues of fundamental importance relating to Canada's emissions-reductions targets: (1) the key role of the provinces and (2) the identification of sectoral priorities. As to the first, the report says (72–3) that, despite the recent Supreme Court decision, "the responsibility for delivering the reductions in line with Canada's 2030 and 2050 objectives rests largely in the constitutional realm of the provinces" and therefore "the federal government remains dependent on the provinces to achieve its most ambitious climate targets." This sobering reminder of reality is compounded by the result of the report's sectoral analysis. In this context, the Trottier Institute report modifies significantly the conclusions reached in other analyses.

Mark Jaccard, for example, has argued persuasively in his book *The Citizen's Guide to Climate Success* that the transportation and electricity sectors are those deserving of special focus in the near-term. These two are in fact intrinsically related, since the solution for the transportation sector is electrification. But the challenge is to continue reducing GHGs in the electricity sector while at the same time greatly expanding its capacity to electrify the transportation and other sectors. Assuming we meet our 2030 target, we are required to move then from 400 to 450 Mt of GHG emissions to zero in twenty years, and to do so we will need a good deal more of non-GHG-emitting electric power. The *Healthy Environment and Healthy Economy* document notes that by 2050 "Canada will need to produce up to two to three times as much clean power as it does now." (A similar projection has been made for the United States; Princeton 2021.) Since Canada is already coming close to utilizing its available suite of hydroelectric installations, the obvious places to expand capacity are in wind and solar renewables. However, as the Trottier Institute report emphasizes, a substantial fraction of the effort needed in the electricity sector must be devoted to building a vast new infrastructure for the delivery of this large new supply of electrical power – and that much new powerline construction will almost certainly engender its own forms of popular opposition (e.g., Gelles 2022).

Furthermore, I for one seriously doubt that Canada can *double or triple* its production of "clean" electrical power by wind and solar renewables alone. Thus it seems to me that serious consideration must be given to promoting the new, small-scale Generation IV nuclear reactors, which are considered to be operationally very safe and which – unlike the older large installations – do not need to be located on lakes, since they use different coolant systems (molten salt rather than water). Numbers of these small-modular-reactor (SMR) plants could be situated near our urban areas. There is already at least one Canadian firm, Terrestrial Energy, which is developing this technology and which received a $20 million investment from the federal government in October 2020. In early December 2021, Ontario Power Generation announced that it had commissioned a US firm, GE Hitachi, to build a more traditional type of SMR (a light-water reactor) at its Darlington station (McClearn 2021). Indeed, the federal government's "SMR Action Plan" treats advances in nuclear-energy technology as an important part of a zero-emissions objective for the electricity sector. However, the Trottier Institute report wisely cautions us that many uncertainties lie ahead in this area, both in terms of successful technological development, as well as in public and regulatory acceptance.

This important report goes on to identify Canada's key sectors in terms of emissions reductions as follows (57; see also the graphic in RBC 2021, 4):

[T]he GHG share of the oil, gas and refining industries has grown systematically over the last 30 years, climbing from 15.7% in 1990 to 19.3% in 2005 and reaching 23.6% of the total in 2019. Despite technological improvements in oil sands production ... this sector contributes close to one-third of energy-related emissions. A similar pattern is observed for the transport sector, whose share of emissions grew unabatedly from 24.1% in 1990 to 25.7% in 2005, and to 29.7% of the total in 2019. Together these two sectors are responsible for more than half the country's emissions; they also underwent the most rapid increase in absolute terms over the entire 1990-2019 period.

To be sure, transportation is the sector where a genuine revolution is in the offing – *although how quickly some radically new technologies can take over the vehicular marketplace is another matter entirely*. All major automobile manufacturers are already offering many different sizes and models of both electric vehicles (EV) and hybrids, or will soon be doing so. Since most auto trips are local, the hybrids can also be used in electric mode only for most of the time. Technologies include hybrid electric vehicles (gas-electric, no plug-in required); plug-in hybrid electrics, permitting longer-range electric use; and hydrogen-fuel-cell electrics, with a combination of fuel cell and electric motors, and water as the only by-product. (Power-hungry consumers concerned about losing their racing fantasies will be blessed with alarming 0-to-100 kph acceleration and simulated internal-combustion-engine sounds in the more expensive EV models.)

In late June 2021, Canada announced that all new light-duty car and passenger truck sales must be zero emission by 2035, and that the federal government would be supporting the development of a national electrical-vehicle charging-station infrastructure. Second, a number of manufacturers are also now producing electric delivery vans and pickup trucks. In January 2021, General Motors announced that it will begin assembling its EV600 delivery van at its Ingersoll, Ontario, plant, an investment sparked by a large advance order from FedEx. Third, firms such as Daimler, Volvo, Tesla, and others will be bringing out both short-haul and long-haul electric-powered tractor-trailer units. Daimler's will go into production in 2021. Also in January 2021, General Motors, the fifth-largest automaker in the world, announced that it will make only electric vehicles by 2035; Volvo has followed suit.

However, the Trottier Institute study argues that transport "does not transform as quickly as might be expected" and electrification there will have almost no impact on meeting GHG targets by 2030, with particular challenges in freight and public transport and little or no impact in the aviation sector. The conclusions to its analysis are clear and direct (208):

> The most cost optimal way to reach 2030 targets is to significantly reduce emissions from the oil and gas sector ... In addition to oil and gas, the industrial, commercial and electricity sectors must bear the largest efforts early on. Governments should therefore focus a major share of their attention on these sectors. Due to the nature of Canada's economy, less than 20% of all GHG emissions can be directly assigned to citizens' direct choices.

One of the major explanations for this perspective is that the transportation sector responds to choices made by millions of consumers, whereas the industrial sector operates within the framework of corporate decision-making by some hundreds of large producers.

During the last two decades, little change in high energy intensity, coupled with rising production in the oil and gas sector – crude oil production more than doubled from 1999 to 2019 – has been a principal factor in Canada's failure to make any reductions in its GHG emissions. The main "takeway" from the Trottier Institute is a harsh one (101): "Canada must reduce its crude oil and natural gas production rapidly before 2030 to conform to net-zero targets" (this sentence is bolded). Thus, in the Trottier Institute report, Canada's industrial sector as a whole, and the oil and gas sector especially, represents by far the most cost-effective and near-term opportunity for emissions reduction. Strategies for this purpose include (183) technological innovation, fuel and/or technology switching, product and process switching, and emissions capture. Carbon capture and storage needs to be greatly expanded in Canada, but it is expensive and is only feasible for very large facilities (that is, it cannot be scaled down).

"Reducing production and demand" in the oil and gas sector, according to the Trottier Institute report, is an inescapable need for Canada in terms of meeting its shorter-term emissions-reductions targets. But this is not what current energy production scenarios have in mind: *The Production Gap Report 2021*, issued by the Stockholm Environmental

Institute and affiliated groups in October (SEI 2021), indicates that the world's group of fossil-fuel-producing countries (including Canada) plans to produce "more than double the amount of fossil fuels in 2030 than would be consistent with limiting global warming to 1.5°C, and 45% more than consistent with limiting warming to 2°C." Citing reports by the quasi-independent federal government agency Canada Energy Regulator (CER 2020), the production-gap report (45) shows that Canada plans to increase oil and gas production by 18 per cent and 17 per cent respectively from 2019 to 2040.

The increases in the oil subsector are destined for export, which allows Canada to take advantage of the accounting scheme under the Paris Agreement whereby a country is responsible only for emissions released on its own territory, but not for emissions released through consumption of its exported products elsewhere. This factor becomes a key part of the "Evolving Policies Scenario" in *Canada's Energy Future 2021* (CER 2021), which assumes "less global demand for fossil fuels, and greater adoption of low-carbon technologies" in Canada and the world as a whole. Whereas Canadian fossil-fuel use is projected to drop by over 40 per cent between 2021 and 2050, Canada's production of crude oil is expected to stay virtually the same (five million barrels per day) during this same period, sustained by the export market. Unsurprisingly, the report concedes that "the Evolving Policies Scenario is unlikely to achieve net-zero emissions by 2050."

As of late 2021, this open contradiction between what is planned and what is needed for Canada to meet its 2030/2050 GHG emissions-reductions targets remains unresolved. Of course, the socio-political implications of the need to reduce production and demand in the oil and gas sector in Canada are both obvious and extremely challenging, given the distribution across provinces of the facilities associated with this sector. To date, the federal government has set country-wide targets and policies that have been only partially coordinated with provincial government actions and policies. *But Canada simply cannot deliver on its international commitments in the absence of a set of seamless, long-term, and ironclad federal-provincial strategies.* All the hardest work in this regard remains to be undertaken.

Conclusion on Mitigation

As of now, fast-track scenarios for Canada's emissions reductions are needed. The country's GHG emissions were almost exactly the same in 2019 as they were in 2005. We have been making promises since 1988, which we have consistently ignored, right up to and including an interim pre-pandemic target for 2020. We have made renewed promises for 2030 and 2050 that, for all practical purposes, we have not yet begun to take seriously, as of 2022, considering the nation as a whole. As stated above, all the hardest work required to realize Canada's commitments for 2030 and 2050 remains to be undertaken. For useful comments on the challenges ahead, see the most recent reports of Canada's commissioner of the environment and sustainable development (CESD 2022) and the interview with Canadian environmental scientist Vaclav Smil in the *New York Times* (Smil 2022).

Over the course of the three decades since Prime Minister Brian Mulroney first committed Canada to a GHG emissions-reduction target at an international forum in 1988, federal-provincial political battles over how to discharge this responsibility have gone on and on. A majority of Canadian provinces, including the largest two, Ontario and Quebec, as well as the oil-producing ones, Alberta and Saskatchewan, spent years fighting the federal government in court over authority to impose a federal carbon tax, an issue that was only finally decided in 2021. As late as 2020, the Alberta government was sponsoring an oddly named "Public Inquiry into Anti-Alberta Energy Campaigns," which was, among other things, offering to the public simplistic "climate denialism" literature.

A change of government at the federal level later in the 2020s could weaken the policies needed to deliver on our current promises (for 2030 and 2050), as occurred once before during the decade from 2006 to 2016. Time is running out for both Canada and the world on the climate-change file. This is why it is imperative that in the coming few years Canada should front-load additional policies and measures in order to securely lock in its commitment to the federal government's interim 2026 target for emissions reductions. I mentioned the idea of front-loading earlier, in recommending that Canada seek to convince its partners in developed nations to provide – via a global decarbonization bond – a full decade's worth of climate finance funding to developing countries by 2023 at the latest. This type of strategy should now be employed on the domestic scene as well. In other words, Canada should implement, over the next four years, a suite of policies, regulations, sector targets, and subsidies that will, at least to some extent, lock in the country on pathways that could reasonably be expected to make significant progress toward our announced GHG emissions-reductions targets.

B. IMPACTS AND ADAPTATION

Among the endless hurdles societies face in crafting a policy response to global warming is this: not only do citizens have to pay for the costs of mitigation, as above, but they must also face the prospect of experiencing adverse climatological impacts as they are doing so. This is especially the case in the Far North, where warming in the Arctic will be considerably worse than it will be in the lower latitudes and where serious problems are already apparent. On the other hand, there is an old saying, "the future will take care of itself," and as we wait, what could be wrong with a little warming for a cold country? The modern age finds comfort in regarding itself as being technologically competent: "When we encounter problems, we find solutions and fix them." The thought that *long-delayed but inevitable – or worse, irreversible – impacts* might cause ruin to the world's civilizations as we know them at some point in the future is just very hard to deal with.

Impacts are usually referred to as the adverse consequences of environmental risks and adaptations are the various adjustments we make to those risks. In an evolutionary sense, adaptation is understood retrospectively, as the emergence in the past of new traits that proved to be successful responses to changing environmental conditions. In the context of climate change today, societies seek to take advantage of scientific knowledge in order to *anticipate* the impacts of new risks that may be detrimental to established ways of life, to devise policies and measures to respond to them, and by so doing to seek to lessen their harmful effects. Where the new risk is a planet-wide phenomenon – as climate change is – distinctions must be made between global impacts, on the one hand, and regional impacts on the other. Here we shall first give a brief summary of the global dimensions, and then focus on Canada, which as a northern-hemisphere nation will experience a particular subset of climate-related risks and therefore face specific challenges in adapting to them.

The following summary of the global dimensions of climate change draws on the elaborate discussion found in one of the main sections of the Fifth Assessment Report prepared by the Intergovernmental Panel on Climate Change (IPCC), *Climate Change 2014: Impacts, Adaptation, and Vulnerability.* These are the major points:

1   Freshwater-related risks: threats to the supply of adequate surface water and groundwater resources, particularly in subtropical regions;

2 Terrestrial and freshwater ecosystems: risk to species from habitat modification, pollution, and invasive species;

3 Coastal systems and low-lying areas: submergence, coastal flooding, and coastal erosion from sea-level rise;

4 Marine systems: threat to fisheries productivity and marine biodiversity from ocean warming and acidification;

5 Food security: threats to the output of major crops (maize, wheat, rice) from temperature increases and drought in both tropical and temperate regions;

6 Urban areas: threats such as heat stress, extreme precipitation, drought, air pollution, water scarcity;

7 Rural areas: impacts from threats to food crops;

8 Human health: impacts of heat stress, poor nutrition, increase in food- and water-borne diseases, changes in insect-borne diseases;

9 Human security: displacement of many people, migrations, violent conflicts, threats to national infrastructures and territorial integrity.

On a large-scale regional level, there are distinctive patterns of risk:

1 Africa: drought, water scarcity, crop productivity, malnutrition, vector- and water-borne diseases;

2 Europe: flooding in river basins and on coasts, heat stress and wildfires, drought and threats to food production;

3 Asia: severe flooding, drought, heat stress, malnutrition;

4 Australasia: flood damage, wildfires, heat stress;

5 North America: wildfires, heat stress, drought, flooding in coastal and riverine areas, extreme weather;

6 Central and South America: decrease in food production, lessened availability of freshwater, heat stress, extreme precipitation, vector-borne diseases;

7 Polar Regions: viability of northern communities and marine ecosystems, infrastructure damage, permafrost melting;

8 The Oceans: decline in fish and vertebrate species, threats to coral reefs and coastal boundary systems, acidification, extreme events.

Climate Change 2022 presents an updated version of these patterns and, for the first time in this series of reports, uses the language of risk in a systematic way to classify the impacts of climate change (see appendix 2).

The Government of Canada issued *Canada's Changing Climate Report* in 2019. In relation to global warming, it pointed specifically to "near-surface and lower-atmosphere air temperature, sea surface temperature, and ocean heat content" as the main risk factors. Near-surface temperature increases are generally higher in the North and the Prairies. It notes that, in order to keep global rises below +2°C, global emissions must "peak almost immediately, with rapid and deep reductions thereafter." Precipitation has been shifting from less snowfall to more rainfall, with a risk of lesser amounts of rainfall in parts of southern Canada, especially the Prairies, if emissions reductions targets are not met. Next, there will be changes in climate extremes: (1) hot temperatures will become more frequent and more intense, with increased risk of drought and wildfires; and (2) inland flooding in urban areas will result from extreme precipitation events. In terms of the cryosphere, western glaciers will lose most of their volume, and warming and thawing of permafrost will have major impacts in the North. The Arctic marine environment will experience extensive ice-free periods during summer. (The organization Berkeley Earth [2021] has useful country-level warming projections to 2100, including for Canada.) The Government of Canada has stated (ECCC 2022): "Canada's average temperatures are rising at twice the global average, and three times in the North."

The oceans will continue to warm, with loss of oxygen and acidification producing adverse impacts on the health of marine ecosystems, especially in the Arctic Ocean. Sea-level rise and local land subsidence will increase the threat of coastal flooding: The latest study (Sweet et al. 2022) predicts a one-foot rise in ocean levels along US coastlines – and thus for Canada as well – by 2050. Reports in late 2020 and 2021 from the US National Oceanic and Atmospheric Administration (NOAA), *Arctic Report Card 2020* and *Arctic Report Card 2021*, focused on this region of greatest vulnerability for all the northern nations, including Canada, whose territories ring the North Pole. Significant new findings include the unusually warm surface air temperatures (3 to 5°C above average) in Siberia during the first half of 2020 and three extreme melt episodes on the Greenland Ice Sheet (as well as an episode of rainfall never before observed) in July-August 2021. There are steadily declining trends in sea ice cover in the summer minimum and winter maximum, warming trends in sea surface temperature, and increasing permafrost thaw (Arctic Institute 2021). In settled areas of the Arctic in Canada, there are major threats to infrastructure, including building foundations, roads, pipelines, and communications networks.

However, a paper by Marshall Burke of Stanford University and others, published in *Nature* in 2015, argued that certain northern-hemisphere nations, including Russia, Canada, and Scandinavia, will have little or no risk of experiencing negative net *economic* impacts from climate change out to 2100. On the contrary, all of them are likely to have a significant net economic benefit. One specific activity in the northern hemisphere could be particularly important: agriculture. Warmer temperatures and longer growing seasons, which are already in evidence (described in a *New York Times Magazine* essay and pictures from December 2020), could greatly enhance food production in the North and parts of Europe, especially in Russia's huge landmass, through a shift gradually northwards in agriculture. Not surprisingly, the economic outlook for most of the southern hemisphere is grim.

Given the clear likelihood that farm yields in places like the United States, South America, India, and Africa will likely decline in a warmer world, with expected widespread food shortages in the poorer nations, the rosier prospect for agriculture in the North will have geopolitical consequences. One might hope that Canada at least would use the promise of a larger food surplus to assist nations around the world whose people will be suffering malnutrition and starvation. On the other hand, even if Canada's agricultural sector should benefit on balance from global warming, the country as a whole will still have to deal with and pay for the other negative aspects of climate change summarized below.

It is worth emphasizing that the Burke study deals only with net national economic forecasts. Thus, even if Canada's economy and agricultural sector as a whole should benefit from warming, the primary locus of farming will shift substantially northwards, whereas some large, more southerly, areas in the West may become unproductive due to persistent drought. This and the expected major expected impacts on our boreal forests (die-back and wildfires), as well as coastal and urban flooding, damage to infrastructures in the North, increasingly violent storms, new insect infestations, heat stress, freshwater supply, and other challenges will be serious indeed. At the same time, there will be perhaps billions of people elsewhere in the world who will be pleading for our help.

How Willing Will Canada Be to Do Its Share to Halt Global Warming?

The suggestion that Canada's economy might benefit from climate change over the longer term prompts us to inquire as to the extent to which a solid majority of Canadian citizens across the land are, as of

2022, fully supportive of their own national government's announced commitments. To repeat, these are threefold: (1) a promise made in 2015 to join with other developed nations to collectively provide, by 2020, $100 billion annually in climate finance to developing countries; (2) specified reductions in GHG emissions by 2030; and (3) the achievement of net-zero emissions by 2050. To its credit, Canada's participation in the collective efforts to rein in emissions through international agreements has been fully supportive of the principle of "fairness" as it applies to the differential responsibilities of nations, particularly the poorer ones. However, over the years, the results of public-opinion polls indicate a reluctance on the part of many to bear any current new costs for emissions-reductions measures. This tells a somewhat different story. Much remains to be done on this front.

Canadians have indeed sometimes wondered about how the fairness principle might be applied to our own situation. They can, if they wish, easily discover the important facts that Canadians make up about 0.5 per cent of the world's population and now contribute about 1.5 per cent of global GHG emissions, two indicators differing by a factor of three. They tend to react to this information in two ways. Either they observe that this is a very small share of the global problem, especially compared with the largest emitters, China and the United States; or they call attention to the fact that we live in a vast, cold country, on the second-largest landmass on the planet, and so naturally require more energy (and its emissions) to stay well and healthy. On the other hand, although we are a small country by population size, Canada is the eleventh largest emitter among all nations, and, on a per-capita basis, we are in second place among the G20 group of nations, behind only Australia (Olivier and Peters 2020, table B.5). What are we to make of these numbers? How does the principle of fairness apply to us? Should we properly have lower reductions targets in view of our special circumstances?

As we proceed throughout the decade of the 2020s on the rough road toward our 2030 target, we need to have a frank discussion on these questions.

We can start with the observation about how small our total emissions are compared with some others, although this complaint is easily dealt with. Since we are the eleventh-largest emitter, many countries rank below us, and many of them toward the bottom have less than our 1.5 per cent of the total. No collective action under the Paris Agreement could possibly succeed if all the others, living in somewhat different circumstances, made the same urgent plea about the relative insignificance of their emissions. In response to the complaint about our cold climate and need for more energy, many other nations, such as India, could point to their need for substantially increased air conditioning, especially as more extreme heat looms. Many of them also have challenges relating to the distances faced by internal transportation. Therefore, neither the alleged insignificance of our emissions, nor the special pleading about our latitude, gains any traction, because some variation of the same case could be made by so many others.

Worse, there is a good argument on the other side which is rarely mentioned: This concerns cumulative – also called "historical" – emissions. Nations that started the process of industrialization long ago (beginning in the nineteenth century) have had much more time to add large amounts of CO_2 emissions to the atmosphere. This has become a serious factor in global warming as a result of the so-called "residence time" of CO_2 in the atmosphere, about one-third of which remains aloft a century after being emitted. Hannah Ritchie's analytical article, "Who Has Contributed Most to Global CO_2 Emissions?" gives estimates for cumulative emissions as of 2017 for the entire period 1751 to 2017, expressed as a share of the world total, as follows: United States, 25 per cent; Europe (EU-28), 22 per cent; China, 13 per cent; Russia, 7 per cent; India, 3 per cent; Canada, 2 per cent.

There are some interesting intermediate steps along the way, all revealing the effect of a leadership position in the process of industrialization. In 1882, more than half of the world's emissions came from the United Kingdom alone; until 1950, more than half came from Europe. In 1990, China's share was still only 5.36 per cent and India's was a mere 1.52 per cent. Finally, let us add per-capita CO_2 emissions in 2019 for this same set of nations (in tonnes, rounded): United States, 16 t; Europe (EU-28), 7 t; China, 8 t; Russia, 12 t; India, 2 t; Canada, 16 t.

Considering this data, it is easy to see why any judgment about fairness in the distribution of responsibility for greenhouse-gas reductions depends on what criterion one choses to apply in any comparative study. For Canada there is no one correct answer to the question of what indicator best determines a fair set of objectives for us on greenhouse-gas emissions in comparison with the rest of the world. Perhaps we could fall back on a simpler approach. The 2015 study by Marshall Burke and others, referred to above, includes an interactive map entitled "Economic Impact of Climate Change on the World," which the reader can consult online. Placing one's cursor over various nations and regions of the world reveals the authors' conclusions that the United States will experience a 36 per cent drop in GDP per capita by 2100; for China, the drop will be 42 per cent; Japan, 35 per cent; Australia, 53 per cent; Mexico, 73 per cent. For Brazil, most of Africa, Saudi Arabia, and India, a complete economic collapse is forecast for the end of the present century. On the other hand, for 2100 Russia is shown at an expected and astonishing growth of 419 per cent GDP per capita; Norway, Sweden, and Finland are over 200 per cent; and the United Kingdom, Germany, Eastern Europe, and Ukraine all are listed with substantial positive forecasts between 40 per cent and 90 per cent. This same study predicts that Canadians will reach the end of this century with a forecasted increase of 247 per cent GDP per capita and facing less than a 1 per cent chance that climate change might reduce our GDP per capita at all.

These are estimates published in 2015 but which extend all the way out to 2100, so we can expect that many additional studies will be done along these lines. These may modify the picture presented there. However, the general conclusion about the radical inequality in the impacts of climate change on economic performance is unlikely to change. A 2019 publication by Noah Diffenbaugh and Marshall Burke summarizes this point as follows: "We find very high likelihood that anthropogenic climate forcing has increased economic inequality between countries ... The primary driver is the parabolic relationship between temperature and economic growth, with warming increasing growth in cool countries and decreasing growth in warm countries." Thus the sounds of any celebrations about this good news among people in a small group of nations in the northern hemisphere, including Canada, will likely be drowned out in the cries of utter misery from the less-fortunate areas. If this forecast turns out to be accurate, billions of people will suffer and very many will die as the global catastrophe plays out over countless decades.

Canadians would be well advised not to revel in any positive forecast for their own future. It would be seemlier if we were to just quietly count our blessings and announce to the rest of the world that we will most assuredly fulfill all three of the commitments we have made to combat climate change, including our pledge of reaching net-zero greenhouse-gas emissions by 2050. And there is another reason why we Canadians should be more focused on near-term objectives than on our country's very long-term prospects with respect to climate change. As noted earlier, so far we have fulfilled none of the pledges on GHG emissions reductions we have made to date.

The Government of Canada stated in late March 2022 (ECCC 2022):

The Canadian Net-Zero Emissions Accountability Act specifies that Canada must establish a 2026 interim GHG objective. Based on Canada's current emissions reduction trajectory, Canada's 2026 interim objective will be 20% below 2005 levels by 2026 [i.e., 584 Mt]. This interim objective is not an official target akin to the 2030 NDC, but progress towards achieving this target will be a cornerstone of future progress reports associated with this 2030 ERP [Emissions Reduction Plan] in 2023, 2025, and 2027.

This is the first time we have had a target as short as four years into the future. Assuming that, when the certified 2022 totals are released in 2023, we are still at or near ~730 Mt emissions, we will have to realize an average annual reduction of 36.5 Mt for each of the coming four years.

We should pay close attention to the promise we have made for 2026 beginning in 2023. The clock is ticking.

Reminiscences and Acknowledgments

I would like to close this discussion by relating a set of personal experiences. During the years 1994 to 2005, I had the great good fortune to hold two externally funded research chairs, the first at Queen's University and the second at the University of Calgary. The chair funders numbered, along with the federal granting councils, large Canadian companies in the chemicals and oil-and-gas sectors. Both chair projects dealt with environmental issues, including climate change, from the perspective of risk management and communication. I can recall many conversations with senior office-holders in both industry sectors, representing both the Canadian Chemical Producers Association (CCPA) and the Canadian Association of Petroleum Producers (CAPP). I held many somewhat testy meetings with these senior industry employees, many of whom were promoters of climate-science denialism, although none of them seemed to have the slightest acquaintance with the scientific literature on the subject, including the IPCC's *Third Assessment Report*, which came out in 2001. I also recall seeing a public notice of an event in Calgary at which two unprepossessing members of the UK's House of Lords were paid by Alberta companies to fly over from England to dispense some utter nonsense about climate change, despite their complete lack of scientific credentials. Meanwhile, I was writing articles for the *Calgary Herald* in 2002, advocating ratification of the Kyoto Protocol. Among the major oil companies, the uninformed rhetoric has as of now mostly subsided. Steve Williams, president and CEO of Suncor, Canada's largest oil company, stated at a public meeting in Calgary in 2018: "Climate change is science. Hardcore science."

☙

After my abrupt and involuntary departure from my position as a full professor in the School of Policy Studies at Queen's University in 2005, a result of the policy of mandatory retirement then in force, I found a new home for my academic research work in risk management at the McLaughlin Centre for Population Health Risk Assessment at the University of Ottawa. The director of the centre, Dan Krewski, and I have collaborated on research projects for some thirty-five years, beginning when he was a civil servant at Health Canada. He and two of our colleagues, Mike Tyshenko and Patricia Larkin, collaborated with me on a major open-access journal article, "Treaty Framing and Climate Science: Challenges in Managing the Risks of Global Warming," published in early 2021 after some three years of effort. I am most grateful to my co-authors for allowing me to reproduce parts of that paper in chapters 5, 6, and 7 of this book.

I am also pleased to return to McGill-Queen's University Press and to editor Philip Cercone and his colleagues with this title, including the honour of being included in a new series at the press, Canadian Essentials, directed by series editor Daniel Béland.

Notes and Calculations for Table 6.4

O&P = Olivier and Peters, *Trends in Global CO₂ and Total Greenhouse Gas Emissions* (2020), table B.1; CAT = *Climate Action Tracker*, 2030 projections as of 15 Sept. 2021 (cf. NewClimate Institute 2021).

CHINA

Average annual percentage change since 2015 in CO_2 emissions = +2.0 per cent (EDGAR 2020); note that China's existing and proposed NDCs refer only to CO_2 emissions, not GHGs. My estimate for 2030 (~16 Gt) assumes a 1.5% growth in GHG emissions per annum for six years (2020 to 2025) and 1 per cent per annum from 2026 to 2030. China's CO_2 emissions grew 1.5 per cent in 2020 and are projected to increase by another 5.5 per cent in 2021 (Jackson et al. 2021). The CAT estimated range is 13.2 to 14.5 GtCO2e per year in 2030, which I believe is much too low: Since the midpoint of that range (~14 Gt) is identical to the actual 2019 figure (in O&P), this suggests that China's GHG emissions have already peaked, which is extremely unlikely. (China has pledged to arrive at peak emissions "before 2030.") China has committed to "phase-down" coal consumption only beginning in 2026, and in 2020 China commissioned a significant number of new coal-fired energy plants (*Global Energy Monitor*, "China Dominates 2020 Coal Plant Development"), which normally have a useful life of around forty years. See also IEA (2021d), Bloomberg (2021), and "A Note on China and Climate Change" at the end of this appendix.

INDIA

Its INDC stated that there was a commitment to almost quadruple its GDP in the sixteen years between 2014 and 2030, with a promise to reduce emissions intensity of GDP by 34 per cent below 2005 levels by 2030. This seems to be a very unrealistic scenario. India's GDP grew by 40 per cent (8 per cent per year) in the five years from 2014 to 2019; one estimate predicts considerably stronger growth from 2020 to 2030, for a total 250 per cent growth from 2014 to 2030 (https://www.statista.com/statistics/263771/gross-domestic-product gdp-in-india/), although this may be on the high side. I have conservatively assumed that an annual growth rate of 8 per cent will continue to 2030, without any major achievements in emissions-intensity reductions; thus, I have increased the GHG emissions figure from 2014 (O&P: 3.3 Gt) to a 2030 estimate of 4.6 Gt. The CAT estimate is 3.84 to 4.02 $GtCO_2e$ in 2030, which I think is a bit too low, given India's coal pipeline, but the uncertainties in India's case are undoubtedly quite high. See SEI (2021, 44) for a future coal estimate.

UNITED STATES

The Paris Pledge (−27 per cent of 1990 levels by 2030) was made in 2015 during the Obama administration. Then, of course, Trump withdrew from Paris. In April 2021, the Biden administration doubled the Obama commitment, now −50 per cent to −52 per cent of 1990 levels by 2030. Since this is such a dramatic step, one does not know how realistic it is, especially because there is a possibility that a Trump administration may return in 2024 – whereupon the United States may once again withdraw from the Paris Agreement. But since the steadily declining energy intensity of the US economy is the principal factor in the ability of the United States to reduce emissions, it is possible that the 2015 commitment level can be achieved irrespective of presidential-election politics, so that number (5.2 Gt) is retained in table 6.4. The CAT estimate is considerably higher (from 6.1 to 6.2 $GtCO_2e$/year by 2030), mainly because the Biden proposals still have to be approved by congressional actions before being implemented, with considerable uncertainties.

RUSSIA

A 2020 NDC commitment of 30 per cent below 1990 levels by 2030 is implausible; I have estimated no change for the period 2019 to 2030, but an increase is just as likely.

Notes and Calculations for Table 6.4

INDONESIA

The 2020 NDC "unconditional" commitment gives 29 per cent below BAU levels (2869 GtCO$_2$e/year) by 2030. The lower "conditional" commitment is implausible.

BRAZIL

An ostensible 2020 NDC commitment of 43 per cent below 2005 levels by 2030 is implausible. I have estimated a 25 per cent increase for the period 2019 to 2030.

MEXICO

An ostensible 2020 NDC commitment of 22 per cent below 2005 levels by 2030 is implausible. I have estimated a 25 per cent increase for the period 2019 to 2030.

CANADA

The enhanced 2030 NDC target of 40 to 45 per cent below 2005 levels by 2030 is very aggressive and contrasts with the country's poor record of meeting its stated commitments to date; I have accepted the CAT 2021 projection.

OTHER

In the cases where there are no reductions commitments in the NDCs, I have estimated (a) a 50 per cent increase for the two major oil-producing nations (Iran and Saudi Arabia) and (b) a 25 per cent increase for the rest (Brazil, Mexico, Turkey, and South Africa), where development pressures will be strong.

IMPLICATIONS OF THE GHG EMISSIONS PROJECTIONS FOR 2030

Climate Action Tracker, a web-based analytical resource, is a joint venture between Climate Analytics, a non-profit climate-science and policy institute based in Berlin, and NewClimate Institute, a non-profit institute established in 2014, in collaboration with the Potsdam Institute for

Climate Impact Research. The "Ratings" provided by *Climate Action Tracker* for the top sixteen countries in table 6.2 and some others compare national policies and GHG emissions with the objectives of the Paris Agreement (https://climateactiontracker.org/countries/). They include the categories of *critically insufficient* ("a country's climate policies and commitments reflect minimal to no action and are not at all consistent with the Paris Agreement") and *highly insufficient* ("indicates that a country's climate policies and commitments are not consistent" with the "+1.5°C target of the Paris Agreement"). A designation of "highly insufficient" is equivalent to a global warming level of 3°C to 4°C; for "critically insufficient," above 4°C. "Insufficient" means that a country's efforts still need improvement. These are the ratings as of 15 September 2021:

China	Highly Insufficient	Mexico	Highly Insufficient
India	Highly Insufficient	Turkey	Not Assessed
USA	Insufficient	S. Africa	Insufficient
EU	Insufficient	Australia	Highly Insufficient
Russia	**Critically Insufficient**	S. Korea	Highly Insufficient
Indonesia	Highly Insufficient	Canada	Highly Insufficient
Brazil	Highly Insufficient		
Iran	**Critically Insufficient**	Argentina	Highly Insufficient
Japan	Insufficient	Kazakhstan	Highly Insufficient
S. Arabia	**Critically Insufficient**	UAE	Highly Insufficient

A NOTE ON CHINA AND CLIMATE CHANGE

The year 2019 marked an ominous milestone in the global emissions of greenhouse gases (GHGs). This is the year when China's emissions first exceeded those of all the world's developed economies combined – that is, of all the members of the OECD plus all of the EU (Rhodium Group 2021). Also, China's level of per-capita GHG emissions, at about ten tonnes per person, has tripled in the past two decades and in 2019 was just below the *average* OECD level (which, however, is much lower than the top level among developed nations, represented by countries such as the United States at 20 t and Canada at about 22 t). Current trend lines indicate that China will likely begin to surpass the OECD average per-capita levels of GHG emissions in the next few years.

China is currently the world's largest emitter of GHGs, at 27 per cent Mt CO_2e of the world's total, followed by the United States at 11 per cent; China represents 31 per cent of global emissions of carbon dioxide, the most important GHG. Finally, in terms of one other key indicator, namely historical cumulative emissions since 1750, China currently sits at 14 per cent, while the developed world together accounts for 51 per cent. Like other countries such as Canada, China will face significant challenges in adhering to the promises it has made for both 2030 and either 2050 or 2060. In fact, those challenges are more severe for China than they will be for the set of developed countries, because in those countries, on the whole, GHG emissions have either stabilized or been declining during recent years, whereas they are still rising in China, and almost certainly will continue to do so for at least another five years. This means that by 2030 China's relative share of global GHG emissions will have increased significantly. On the other hand, China has promised to ensure that its emissions have peaked "before 2030" and that the country will arrive at the objective of net-zero emissions by 2060 (a decade later than many other countries that have promised to arrive at net-zero by 2050).

Important analysis from the International Energy Agency shows that China occupies a distinctive middle position between advanced economies (AEs), on the one hand, and the large group of emerging market and developing economies (EMDEs), on the other (IEA 2021e, 223). In other words, China is neither an AE nor an EMDE, but rather has aspects of both, although it is clearly heading toward the status of an AE. Data such as the World Bank's "GDP per capita" and "CO2e per capita" also illustrate this point, placing China squarely between the sets of developed and developing economies. China has also been, for the last two decades, the most dynamic still-developing economy in the world.

On the other hand, we must remember that, although China is the largest annual emitter *nation* in the world, it is still lower than all developed economies taken together in terms of *per-capita* emissions and far lower in terms of *cumulative* (historical) emissions. No one of these criteria is more important than the other two when it comes to responsibility for the issue of climate change. The COP discussions under the

Paris Agreement tend to focus on nations, rather than individuals or history, because the mechanisms under the Agreement are based on the so-called Nationally Determined Contributions (NDCs) and because nations are the signatories to the Agreement. A small group of those nations are the "major emitters," and they are (in order) China, the United States, the European Union plus the United Kingdom (twenty-eight countries), India, Russia, and Japan. Together, those entities account for two-thirds (67 per cent) of all carbon-dioxide-gas emissions in the entire world. I believe that all six *collectively*, not China alone, have the major responsibility for addressing the issue of climate change. *China in no sense bears more responsibility than anyone else for creating the problem of climate change!*

In a very real sense, China is facing many of the world's dilemmas with respect to the need to stabilize and reduce GHG emissions. One of those dilemmas is that all EMDEs must continue to grow their economies and their GHG emissions, even as AEs generally reduce their own emissions, and China, with its recent success in this regard, could be said to best understand that dilemma. China certainly also understands that all EMDEs need a great deal of financial assistance to transition to non-polluting energy sources. China is therefore also uniquely positioned to assume global leadership on using the Paris Agreement and United Nations frameworks to address the issue of climate change. The European Union likely would support China in this regard, but cannot itself be the global leader, since it does not have a close historical connection to the EMDEs. On the other hand, no one can count on the United States to exert any real leadership on this issue: for twenty-five years, ever since 1997, when the US Senate refused to even consider ratifying the Kyoto Protocol, the national politics of that country have consistently frustrated global attempts to deal effectively with the issue of climate change. This leaves only China to do the job.

Given the unsurpassed importance of this issue in coming decades, should China take up the challenge of leadership on climate change, openly and proudly in the United Nations General Assembly, a tectonic shift for the better in international relations might occur.

The Risk Approach in IPCC's AR6 –
Impacts of Climate Change

Climate Change 2022: Impacts, Adaptation, and Vulnerability
Chapter 16:
"Key Risks across Sectors and Regions" [PDF file pages 2766ff.]
Chapter 17:
"Decision Making Options for Managing Risks" [PDF file pages 2939ff.]

Note: Readers who wish to have a full understanding of the risks of climate change are encouraged to study these two chapters, especially the pages and sections listed below.

Chapter 16, "Key Risks across Sectors and Regions"

1   "We particularly look at a set of eight 'representative key risks' that exemplify the 30 underlying sets of key risks identified in the earlier chapters: risk to the integrity of low-lying coastal socio-ecological systems, risk to terrestrial and ocean ecosystems, risk to critical physical infrastructure and networks, risk to living standards (including economic impacts, poverty and inequality), risk to human health, risk to food security, risk to water security, and risk to peace and mobility" (chap. 16, 12).

2   "A key risk is defined as a potentially severe risk and therefore especially relevant to the interpretation of dangerous anthropogenic interference (DAI) with the climate system, the prevention of which is the ultimate objective of the UNFCCC as stated in its Article 2" (chap 16, 56).

3   16.5.2.3 Assessment of Representative Key Risks (chap. 16, 60–80). Each of the eight key risks listed above are explicated in some detail.

4   Solar Radiation Modification (chap. 16, 83–9).

5   Estimating Global Economic Impacts from Climate Change (chap. 16, 111–16).

6   Tipping Points or "large-scale singular events" (chap. 16, 116–19).

7   Summary (chap. 16, 119–20).

8   Alternatively, readers may also go directly here (chap. 16, 121–7, PDF file page 2886), to "Frequently Asked Questions," for a convenient summary of the main points:

FAQ 16.1: What are key risks in relation to climate change?
FAQ 16.2: How does adaptation help to manage key risks and what are its limits?
FAQ 16.3: How do climate scientists differentiate between impacts of climate change and changes in natural or human systems that occur for other reasons?
FAQ 16.4: What adaptation-related responses to climate change have already been observed, and do they help reduce climate risk?
FAQ 16.5: How does climate risk vary with temperature?

9   Also, for another convenient summary, check the *Summary for Policymakers* of *Climate Change 2022: Impacts, Adaptation and Vulnerability,* pages SPM 18–20, for the sections entitled "Complex, Compound and Cascading Risks" and "Impacts of Temporary Overshoot."

Chapter 17, "Decision Making Options for Managing Risks"

1   Executive Summary (chap. 17, 3–7).

2   Frequently Asked Questions (chap. 17, 103–9):

FAQ 17.1: Which guidelines, instruments and resources are available for decision-makers to recognize climate risks and decide on the best course of action?
FAQ 17.2: What financing options are available to support adaptation and climate resilience?
FAQ 17.3: Why is adaptation planning along a spectrum from

incremental to transformational adaptation important in a warming world?

FAQ 17.4: Given the existing state of adaptation, and the remaining risks that are not being managed, who bears the burden of these residual risks around the world?

FAQ 17.5: How do we know whether adaptation is successful?

Supplemental Resources:

IPCC, *Climate Change 2021: The Physical Science Basis,* "FAQs"
IPCC, *Climate Change 2022, Mitigation of Climate Change,* Press Release
IPCC, *Climate Change 2022: Synthesis Report* (expected October 2022)

References and Sources

Only a limited number of references are listed here; the reader can find many more resources for the topics covered in this book on the World Wide Web by putting the relevant headings into a search engine. In doing so, one must examine carefully the sources and attributions for any of the information found there and evaluate them critically as to their trustworthiness. In order to facilitate the reader's further information search, I have included below the specific URLs for a considerable number of valuable academic journal articles which are available in their entirety to the public. All URLs were accessed on 6 June 2022.

The website *Our World in Data* has an elaborate section called "CO_2 Greenhouse Gas Emissions" (https://ourworldindata.org/co2-and-other-greenhouse-gas-emissions). One can find there many fine graphics, especially interactive data visualization charts and maps, and much useful explanatory material. The director of this project, Max Roser, is affiliated with the Oxford Martin School, a research and policy unit based in the Social Sciences Division at the University of Oxford. It is widely regarded as a credible and reliable source for information on many topics of global importance. Another reliable website is *Carbon Brief* (https://www.carbonbrief.org/about-us), a UK-based group financed by the European Climate Foundation, which has superb interpretative essays and data visualization resources.

By far the best single short book written by a climate scientist for the general public is Richard Alley's *The Two-Mile Time Machine* (see chapter 2 references). And for a recent broad overview of the current state of climate science, as well as prospects for reaching the ambitious greenhouse-gas emissions reductions targets that are now accepted by governments in all advanced economies (including Canada), see Zeke Hausfather,

"Written Testimony for the U.S. House of Representatives" (12 March 2021), 36 pages: http://berkeleyearth.org/wp-content/uploads/2021/03/Zeke-Hausfather-Congressional-Testimony-highrez.pdf.

If the reader wishes to examine just one comprehensive and highly credible study that summarizes the scientific evidence on climate change, I recommend this one: United States Global Change Research Program, *Climate Science Special Report* (CSSR, 2017), 470 pages: http://science2017.globalchange.gov/. Given its length, one might wish to dip into it gradually over time; among other things, one will discover that it is very well written. "Highlights" and the "Executive Summary" will be found on pages 10 to 34. In addition, each chapter begins with a section called "Key Findings."

Another indispensable resource for the reader is chapter 19, "Climate Change," in the Canadian textbook *Physical Geology*, by Steven Earle and Karla Panchuk; the entire book is available on the Internet (https://opentextbc.ca/physicalgeology2ed/). At the beginning of chapter 19, there is a marvellous picture (figure 19.0.1) of a core sample from an ocean drilling project which illustrates how scientists interpret evidence of past climate, in this case the hundred-thousand-year warm period known as the "Paleocene–Eocene Thermal Maximum" beginning 55.7 million years ago. This and the remainder of that short chapter are essential for understanding climate and climate change. Chapter 4, "Volcanism," as well as chapter 8, titled "Measuring Geologic Time," and chapter 16, "Glaciation," are also valuable.

One of the most important purposes of this book is to persuade the reader that she or he can trust the strong consensus position on climate change that has been articulated by world climate scientists. My approach is to encourage citizens who, like me, are not expert in the natural sciences just to try to understand *how* scientists go about their work, whether on climate or any other subject. Here I have selected a 2021 article from a reputable science-oriented magazine, demonstrating the process of scientific experimentation, evidence-gathering, and continuous updating with new information. The research area is related to the ongoing effort to integrate all of the many and complex elements that make up the understanding of climate: Max Koslov, "Cloud-Making Aerosol Could Devastate Polar Sea Ice," *Quanta Magazine*, 23 February 2021: https://www.quantamagazine.org/cloud-making-aerosol-could-devastate-polar-sea-ice-20210223/

Articles in *Quanta Magazine*, published by the Simons Foundation, cited just above and in the references to chapters 1 and 2, are in my opinion the best science journalism I know for explaining complex scientific ideas accurately and clearly in understandable language and with useful illustrations.

The URLs for the references below may be difficult to copy into a browser if one is using the print edition of this book. In such a case it is usually possible to locate the full text quickly by entering the author's name and the title (for journal articles) or just the title for reports and documents. Every item in the list of references is freely available on the Internet for all readers, with the exception of most of the books and a few articles.

GENERAL

Jaccard, Mark. 2020. *The Citizen's Guide to Climate Success*. Cambridge: Cambridge University Press.

Leiss, William, and Stephen Hill. 2004. "A Night at the Climate Casino: Canada and the Kyoto Quagmire." Chapter 10 in William Leiss and Douglas Powell, *Mad Cows and Mother's Milk*. 2nd ed. McGill-Queen's University Press.

Simpson, Jeffrey, Mark Jaccard, and Nic Rivers. 2007. *Hot Air: Meeting Canada's Climate Change Challenge*. Toronto: McClelland & Stewart.

Weaver, Andrew. 2008. *Keeping our Cool: Canada in a Warming World*. Toronto: Penguin Canada.

RECENT REPORTS FROM THE INTERGOVERNMENTAL PANEL ON CLIMATE CHANGE (IPCC)

Intergovernmental Panel on Climate Change (IPCC), *Synthesis Report of the Sixth Assessment Report*, expected release October 2022: https://www.ipcc.ch/ar6-syr/.

Climate Change 2022: Mitigation of Climate Change:

Intergovernmental Panel on Climate Change (IPCC). 2022. *Climate Change 2022: Mitigation of Climate Change*. Working Group III contribution to the Sixth Assessment Report of the Intergovernmental Panel on Climate Change. https://report.ipcc.ch/ar6wg3/pdf/IPCC_AR6_WGIII_FinalDraft_FullReport.pdf.

– 2022. *Climate Change 2022: Mitigation of Climate Change: Summary for Policymakers.* https://report.ipcc.ch/ar6wg3/pdf/IPCC_AR6_WGIII_SummaryForPolicymakers.pdf.

– 2022. *Climate Change 2022: Mitigation of Climate Change, Summary for Policymakers,* figure SPM.4. https://www.ipcc.ch/report/ar6/wg3/figures/summary-for-policymakers/figure-spm-4.

– 2022. *Climate Change 2022: Mitigation of Climate Change.* Press Release. https://report.ipcc.ch/ar6wg3/pdf/IPCC_AR6_WGIII_PressRelease-English.pdf.

Climate Change 2022: Impacts, Adaptation and Vulnerability:

Intergovernmental Panel on Climate Change (IPCC). 2022. *Climate Change 2022: Impacts, Adaptation and Vulnerability,* Working Group II contribution to the Sixth Assessment Report of the Intergovernmental Panel on Climate Change. https://report.ipcc.ch/ar6wg2/pdf/IPCC_AR6_WGII_FinalDraft_FullReport.pdf.

– 2022. *Climate Change 2022: Impacts, Adaptation and Vulnerability: Summary for Policymakers.* https://report.ipcc.ch/ar6wg2/pdf/IPCC_AR6_WGII_SummaryForPolicymakers.pdf.

– 2022. *Climate Change 2022: Impacts, Adaptation and Vulnerability: Press Conference Slides.* https://report.ipcc.ch/ar6wg2/pdf/IPCC_AR6_WGII_PressConferenceSlides.pdf.

Climate Change 2021:
The Physical Science Basis

CarbonBrief. "In-Depth Q&A: The IPCC's Sixth Assessment Report on Climate Science." https://www.carbonbrief.org/in-depth-qa-the-ipccs-sixth-assessment-report-on-climate-science.

Intergovernmental Panel on Climate Change (IPCC). 2021. *Climate Change 2021: The Physical Science Basis.* Working Group I contribution to the Sixth Assessment Report of the Intergovernmental Panel on Climate Change. https://www.ipcc.ch/report/ar6/wg1/downloads/report/IPCC_AR6_WGI_Full_Report.pdf.

– 2021. *Climate Change 2021: The Physical Science Basis: Summary for Policymakers.* https://www.ipcc.ch/report/ar6/wg1/downloads/report/IPCC_AR6_WGI_SPM_final.pdf.

– 2021. *Climate Change 2021: The Physical Science Basis, Summary for Policymakers*, figure SPM.1. https://www.ipcc.ch/report/ar6/wg1/figures/summary-for-policymakers/figure-spm-1/.
– 2021. *Climate Change 2021: The Physical Science Basis*. Frequently Asked Questions. https://www.ipcc.ch/report/ar6/wg1/downloads/faqs/IPCC_AR6_WGI_FAQs.pdf.

Other Recent IPCC Reports

Intergovernmental Panel on Climate Change (IPCC). 2019. *Climate Change and Land*. https://www.ipcc.ch/site/assets/uploads/2019/11/SRCCL-Full-Report-Compiled-191128.pdf.
– 2019. *Summary for Policymakers*. https://www.ipcc.ch/srccl/chapter/summary-for-policymakers/.
– 2019. *The Ocean and Cryosphere*. https://www.ipcc.ch/site/assets/uploads/sites/3/2022/03/SROCC_FullReport_FINAL.pdf; https://www.ipcc.ch/srocc/chapter/summary-for-policymakers/; https://climateanalytics.org/media/ipcc-srocc-briefing.pdf.
– 2018. *Global Warming of 1.5°C*. https://www.ipcc.ch/site/assets/uploads/sites/2/2019/06/SR15_Full_Report_High_Res.pdf.
– 2018. *Summary for Policymakers*. https://www.ipcc.ch/sr15/chapter/spm/.

PREFACE

The Debunking Handbook 2020. https://www.climatechangecommunication.org/wp-content/uploads/2020/10/DebunkingHandbook2020.pdf.
"Scientific Method" in the *Stanford Encyclopedia of Philosophy*. https://plato.stanford.edu/entries/scientific-method/.
"Scientific Method" (Wikipedia, accessed 6 June 2022).

INTRODUCTION

Intergovernmental Panel on Climate Change (IPCC). 2020. "The Concept of Risk in the IPCC Sixth Assessment Report: A Summary of Cross-Working Group Discussions." 4 September 2020. https://www.ipcc.ch/site/assets/uploads/2021/01/The-concept-of-risk-in-the-IPCC-Sixth-Assessment-Report.pdf.
Fleming, James Rodger. 1998. *Historical Perspectives on Climate Change*. Oxford: Oxford University Press.

References and Sources to Introduction and Chapter One

Haller, Stephen. 2002. *Apocalypse Soon? Wagering on Warnings of Global Catastrophe*. Montreal and Kingston: McGill-Queen's University Press.

Harvey, L.D. Danny. 2000 [e-book 2016]. *Global Warming*. Milton Park, UK: Routledge.

– 1999 [e-book 2016]. *Climate and Global Environmental Change*. Milton Park, UK: Routledge.

Leiss, William. 2001. *In the Chamber of Risks: Understanding Risk Controversies*. Montreal and Kingston: McGill-Queen's University Press.

Topics and Recommended Websites

"Climate" (Wikipedia, accessed 6 June 2022).

"Evolution of the Sun" (northwestern.edu).

"Large Igneous Provinces" (Wikipedia, accessed 6 June 2022).

"Snowball Earth" (Wikipedia, accessed 6 June 2022).

CHAPTER ONE

Balower, Timothy, and David Bice, College of Earth and Mineral Science, The Pennsylvania State University, in collaboration with the US National Aeronautics and Space Administration. "Earth 103: Earth in the Future" (Creative Commons License). https://www.e-education.psu.edu/earth103/node/508.

Earle, Steven, and Karla Panchuk. 2019. "Climate Change." In *Physical Geology*, 2nd edition, 593–615. BC Open Textbook Collection, available in its entirety at https://opentextbc.ca/physicalgeology2ed/. Chapter 19, "Climate Change," 593–615.

Lee, Howard. 2021. "Scientists Pin Down When Earth's Crust Cracked, Then Came to Life." *Quanta Magazine*, 25 March. https://www.quantamagazine.org/ancient-rocks-reveal-when-earths-plate-tectonics-began-20210325/.

– 2020. "Sudden Ancient Global Warming Event Traced to Magma Flood." *Quanta Magazine*, 19 March. https://www.quantamagazine.org/sudden-ancient-global-warming-event-traced-to-magma-flood-20200319/.

O'Callaghan, Brendan. 2022. "A Solution to the Faint-Sun Paradox Reveals a Narrow Window for Life." *Quanta Magazine*, 27 January. https://www.quantamagazine.org/the-sun-was-dimmer-when-earth-formed-how-did-life-emerge-20220127/.

Witze, Alexandra. 2017. "Ancient Volcanoes Exposed." *Nature* 543 (16 March): 295–6. https://www.nature.com/news/polopoly_fs/1.21630!/menu/main/topColumns/topLeftColumn/pdf/543295a.pdf.

References and Sources to Chapter Two

Topics and Recommended Websites

"Climate" (Wikipedia, accessed 6 June 2022).
"Evolution of the Sun" (northwestern.edu).
"Large Igneous Provinces" (Wikipedia, accessed 6 June 2022).
"Snowball Earth" (Wikipedia, accessed 6 June 2022).

CHAPTER TWO

Alley, Richard B. 2000. "Ice-Core Evidence of Abrupt Climate Changes."
Proceedings of the National Academy of Sciences 97 (4): 1331–4. Open access.
. https://www.pnas.org/content/97/4/1331.
– 2014. *The Two-Mile Time Machine: Ice Cores, Abrupt Climate Change, and
Our Future*. Princeton, NJ: Princeton University Press. Ebook edition avail-
able. https://press.princeton.edu/books/ebook/9781400852246/
the-two-mile-time-machine.
Bender, Michael, et al. 1997. "Gases and Ice Cores." *Proceedings of the National
Academy of Sciences* 94 (16): 8343–9. Open access. https://www.pnas.org/
content/94/16/8343.
Brovkin, Victor, et al. 2021. "Past Abrupt Changes, Tipping Points, and
Cascading Impacts in the Earth System." *Nature Geoscience* 14: 550–8.
https://doi.org/10.1038/s41561-021-00790-5.
CarbonBrief, "Explainer: How the Rise and Fall of CO2 Levels influenced the
Ice Ages." https://www.carbonbrief.org/explainer-how-the-rise-and-fall-of-
co2-levels-influenced-the-ice-ages.
Colose, Chris, et al. 2020. *Energy Education: Climate Forcing*. https://
energyeducation.ca/encyclopedia/Climate_forcing.
Davies, Bethan. 2020. "Ice Core Basics." https://www.antarcticglaciers.org/
glaciers-and-climate/ice-cores/ice-core-basics/.
Jacob, D.E., et al. 2017. "Planktic Foraminifera Form Their Shells via Metastable
Carbonate Phases." Open access. https://www.nature.com/articles/s41467-
017-00955-0.
Lee, Howard. 2020. "How Earth's Climate Changes Naturally (and Why
Things Are Different Now)." *Quanta Magazine*, 21 July. https://www.
quantamagazine.org/how-earths-climate-changes-naturally-and-why-things-
are-different-now-20200721/.
Oregon State University. 2014. "Study Resolves Discrepancy in Greenland
Temperatures during End of Last Ice Age." https://today.oregonstate.edu/
archives/2014/sep/study-resolves-discrepancy-greenland-temperatures-
during-end-last-ice-age.

Steffen, Will, et al. 2015. "Planetary Boundaries: Guiding Human Development on a Changing Planet." *Science* 347. Open access. https://science.sciencemag.org/content/347/6223/1259855/tab-pdf.

Wolchover, Natalie. 2019. "A World without Clouds." *Quanta Magazine*, 25 February.https://www.quantamagazine.org/cloud-loss-could-add-8-degrees-to-global-warming-20190225/.

Topics and Recommended Websites

"Abrupt Climate Change" (NCDC/NOAA).
"Glacial/Interglacial Periods" (NCDC/NOAA).
"Holocene Climatic Optimum" (Wikipedia, accessed 6 June 2022).
"Warm Periods" (climate.gov).

CHAPTER THREE

Aengenheyster, Matthias, et al. 2018. "The Point of No Return for Climate Action: Effects of Climate Uncertainty and Risk Tolerance." *Earth System Dynamics* 9: 1085–95. Open access. https://doi.org/10.5194/esd-9-1085-2018.

Alley, Richard B. et al. N.d. The Pennsylvania State University, Earth 104: Energy and the Environment, Module 5, "The Vostok Ice Core." https://www.e-education.psu.edu/earth104/node/1267 (Creative Commons License).

Arrhenius, Gustav. 1896. "On the Influence of Carbonic Acid in the Air upon the Temperature of the Ground." *Philosophical Magazine and Journal of Science* 5, no. 41: 237–76. https://www.rsc.org/images/Arrhenius1896_tcm18-173546.pdf.

Edwards, Paul N. 2011. "History of Climate Modeling." *WIREs Climate Change* 2, no. 1: 128–39. https://deepblue.lib.umich.edu/bitstream/handle/2027.42/79438/95_ftp.pdf.

EPA. 2022. US Environmental Protection Agency, "Causes of Climate Change," February. https://www.epa.gov/climatechange-science/causes-climate-change.

Fairbank, Viviane. 2021. "Climate Science Is a Fact." *Globe and Mail*, 30 October. https://www.theglobeandmail.com/opinion/article-climate-change-is-a-fact-but-to-prove-it-scientists-are-bogged-down-in/.

"How Do Climate Models work?" (CarbonBrief, 2018). https://www.carbon-brief.org/qa-how-do-climate-models-work.

Intergovernmental Panel on Climate Change (IPCC). 2007. *AR4 Climate Change 2007: The Physical Science Basis*. (Chapter 1: "Historical Overview of

Climate Change Science.") https://www.ipcc.ch/site/assets/uploads/2018/03/ar4-wg1-chapter1.pdf.

– 2021. *Climate Change 2021: The Physical Science Basis. Summary for Policymakers.* https://www.ipcc.ch/report/ar6/wg1/downloads/report/IPCC_AR6_WGI_SPM.pdf.

Lindsey, Rebecca. 2020. "Climate Change: Atmospheric Carbon Dioxide." https://www.climate.gov/news-features/understanding-climate/climate-change-atmospheric-carbon-dioxide.

NAP. 2020. US, National Academies Press, *Climate Change: Evidence and Causes: Update 2020.* https://nap.nationalacademies.org/download/25733.

National Research Council. 1979. *Carbon Dioxide and Climate: A Scientific Assessment.* Washington, DC: The National Academies Press. http://nap.edu/12181.

Revelle, Roger, and Hans E. Suess. 1957. "Carbon Dioxide Exchanges between Atmosphere and Ocean and the Question of an Increase of Atmospheric CO_2 during the Past Decades." *Tellus* 9, no. 1: 18–27. Open access. https://www.tandfonline.com/doi/pdf/10.3402/tellusa.v9i1.9075.

Rodhe, Robert. 2021. "Global Temperature Report 2021." Berkeley Earth. http://berkeleyearth.org/global-temperature-report-for-2021.

Schmidt, Gavin. 2014. "The Emergent Patterns of Climate Change." TED Talk. https://www.ted.com/talks/gavin_schmidt_the_emergent_patterns_of_climate_change.

Semeniuk, Ivan. 2021a. "Humans Are, Beyond Any Reasonable Scientific Doubt, the Primary Cause of Climate Change, UN Report Says." *Globe and Mail*, 9 August. https://www.theglobeandmail.com/canada/article-humans-to-blame-for-acceleration-in-climate-change-report/.

Steffen, Will, et al. 2018. "Trajectories of the Earth System in the Anthropocene" and "Appendix: Supporting Information: Holocene Variability and Anthropocene Rates of Change." *Proceedings of the National Academy of Sciences* 115, no. 33 (14 August), 8252–9. Open access. https://www.pnas.org/content/115/33/8252.

United States. NASA. 2021. "The Causes of Climate Change." https://climate.nasa.gov/causes/.

United States. National Academy of Sciences. 1979. *Carbon Dioxide and Climate: A Scientific Assessment.* Washington, DC: The National Academies Press. http://nap.edu/12181.

Wagner, Gernot, and Richard J. Zeckhauser. 2018. "Confronting Deep and Persistent Climate Uncertainty." Harvard University, HKS Faculty Research Working Paper RWP16-025. https://www.hks.harvard.edu/publications/confronting-deep-and-persistent-climate-uncertainty.

Topics and Recommended Websites

"History of Climate Change Science" (Wikipedia, accessed 6 June 2022).
"Greenhouse Effect" (Wikipedia, accessed 6 June 2022).

CHAPTER FOUR

Anderegg, W.R.L., et al. 2010. "Expert Credibility in Climate Change."
Proceedings of the National Academy of Sciences 107, no. 27 (6 July): 12107–9.
Open access. https://www.pnas.org/content/pnas/107/27/12107.full.pdf.
Bar-On, Y.M., et al. 2018. "The Biomass Distribution on Earth." *Proceedings of
the National Academy of Sciences* 115, no. 25 (19 June): 6506–11. Open access.
https://www.pnas.org/content/pnas/115/25/6506.full.pdf.
Carlton, J.S., et al. 2015. "The Climate Change Consensus Extends beyond
Climate Scientists." *Environmental Research Letters* 10 (2015): 094025. Open
access. https://iopscience.iop.org/article/10.1088/1748-9326/10/9/094025/pdf.
Climate Action Tracker (CAT). 2021. "Glasgow's 2030 Credibility Gap."
November. https://climateactiontracker.org/documents/997/CAT_2021-11-
09_Briefing_Global-Update_Glasgow2030CredibilityGap.pdf.
Cook, John, et al. 2013. "Quantifying the Consensus on Anthropogenic Global
Warming in the Scientific Literature." *Environmental Research Letters*
8 (2013): 024024 (7pp.) Open access. https://iopscience.iop.org/
article/10.1088/1748-9326/8/2/024024/pdf.
– 2016. "Consensus on Consensus: A Synthesis of Consensus Estimates on
Human-Caused Global Warming." *Environmental Research Letters* 11 (2016):
048002. Open access. https://iopscience.iop.org/article/10.1088/1748-9326/
11/4/048002/pdf.
Doran, P.T., and M.K. Zimmerman. 2009. "Examining the Scientific
Consensus on Climate Change." *Eos* 90, no. 3 (20 January). Open access.
https://agupubs.onlinelibrary.wiley.com/doi/epdf/10.1029/2009EO030002.
Leiss, William, and Stephen Hill. 2002. "Why Canada Should Ratify the Kyoto
Protocol." *Calgary Herald*, 11, 12, and 13 April. http://leiss.ca/wp-content/
uploads/2009/12/Why-Canada-should-ratify-Kyoto.pdf.
– "Kyoto Protocol Archive." http://leiss.ca/?page_id=144.
Smil, Vaclav. 2012. *Harvesting the Biosphere*. Cambridge, MA: MIT Press.
Verheggen, Bart, et al. 2014. "Scientists' Views about Attribution of Global
Warming." *Environmental Science and Technology* 48: 8963–71. Open access.
https://pubs.acs.org/doi/pdf/10.1021/es501998e.

Xu, Chi, et al. 2020. "Future of the Human Climate Niche." *Proceedings of the National Academy of Sciences* 117, no. 21 (14 August): 11350–5. Open access. https://www.pnas.org/content/pnas/117/21/11350.full.pdf.

CHAPTER FIVE

Black, Richard, et al. 2021. *Taking Stock: A Global Assessment of Net Zero Targets.* Energy and Climate Intelligence Unit and Oxford Net Zero. https://eciu.net/analysis/reports/2021/taking-stock-assessment-net-zero-targets.

Canada. 2021. Bill C-12, "Net-Zero Emissions Accountability Act." https://www.parl.ca/DocumentViewer/en/43-2/bill/C-12/royal-assent.

– 2021. Environment and Climate Change Canada. *Greenhouse Gas Emissions.* https://www.canada.ca/en/environment-climate-change/services/environmental-indicators/greenhouse-gas-emissions.html.

CarbonBrief. https://www.carbonbrief.org/the-carbon-brief-profile-canada.

Climate Action Tracker. https://climateactiontracker.org/countries/canada/.

Our World in Data. https://ourworldindata.org/co2/country/canada?country=~CAN.

Gattinger, Monica. 2021. "What the IEA's Net Zero by 2050 Report Means for Canada." *Daily Oil Bulletin,* 25 May. https://www.dailyoilbulletin.com/article/2021/5/25/what-the-ieas-net-zero-by-2050-report-means-for-ca/.

Grubb, Michael. 2016. "Full Legal Compliance with the Kyoto Protocol's First Commitment Period – Some Lessons." *Climate Policy* 16, no. 6: 673–81. Open access. https://doi.org/10.1080/14693062.2016.1194005.

International Energy Agency. 2021b. *Net Zero by 2050: A Roadmap for the Global Energy Sector* (May). https://www.iea.org/reports/net-zero-by-2050.

Kyoto Protocol. 2021. https://unfccc.int/resource/docs/convkp/kpeng.pdf and https://unfccc.int/kyoto_protocol.

Leiss, William, Michael Tyshenko, Patricia Larkin, and Daniel Krewski. 2020. "Treaty Framing and Climate Science: Challenges in Managing the Risks of Global Warming." Open access. https://dataverse.harvard.edu/dataset.xhtml?persistentId=doi:10.7910/DVN/2SUI27.

Maciunas, Silvia, and Géraud de Lassus Saint-Geniès. 2018. *The Evolution of Canada's International and Domestic Climate Policy: From Divergence to Consistency?* Open access. https://www.cigionline.org/sites/default/files/documents/Reflections%20Series%20Paper%20no.21%20Maciunas.pdf.

National Round Table on the Environment and the Economy. 2021. "Canada's Emissions Story." http://nrt-trn.ca/chapter-2-canadas-emissions-story.

Pisani-Ferry, Jean. 2021. "Climate Policy Is Macroeconomic Policy, and the Implications Will Be Significant." August. https://www.piie.com/system/files/documents/pb21-20.pdf.

Royal Bank of Canada (RBC). 2021. "The $2 Trillion Transition: Canada's Road to Net-Zero." (October). https://royal-bank-of-canada-2124.docs.contently.com/v/the-2-trillion-transition-canadas-road-to-net-zero-pdf.

Thompson, Helen. 2022. "It's Not Just High Oil Prices, It's a Full-Blown Energy Crisis." *New York Times*, 23 April. https://www.nytimes.com/2022/04/23/opinion/oil-gas-energy-prices-russia-ukraine.html.

Toronto Conference. 1988. "The Changing Atmosphere: Implications for Global Security." Conference statement. http://cmosarchives.ca/History/ChangingAtmosphere1988e.pdf.

Trottier Energy Institute/Institut de l'énergie Trottier. 2021. *Canadian Energy Outlook 2021: Horizon 2060* (October). https://iet.polymtl.ca/wp-content/uploads/delightful-downloads/CanadianEnergyOutlook2021.pdf.

United States Environmental Protection Agency (EPA). 2021. *Inventory of U.S. Greenhouse Gas Emissions and Sinks, 1990–2019.* https://www.epa.gov/ghgemissions/inventory-us-greenhouse-gas-emissions-and-sinks.

United Nations. 2021. *United Nations Framework Convention on Climate Change* [UNFCCC]. https://unfccc.int/resource/docs/convkp/conveng.pdf.

Wright, David V. 2020. "Bill C-12, Canadian Net-Zero Emissions Accountability Act: *A Preliminary Review*" (November 23), online: ABlawg. http://ablawg.ca/wp-content/uploads/2020/11/Blog_DVW_Bill_C12.pdf.

CHAPTER SIX

Canada. Office of the Auditor General. 2018. *Perspectives on Climate Change Action in Canada.* https://www.oag-bvg.gc.ca/internet/English/parl_otp_201803_e_42883.html.

Canada. 2019. *Fourth Biennial Report on Climate Change.* Environment and Climate Change. https://unfccc.int/sites/default/files/resource/br4_final_en.pdf.

– 2020. *National Inventory Report, 1990–2018.* https://unfccc.int/documents/224829.

– 2021. *References re Greenhouse Gas Pollution Pricing Act.* https://decisions.scc-csc.ca/scc-csc/scc-csc/en/item/18781/index.do.

Chalifour, Nathalie J. 2016. "Canadian Climate Federalism: Parliament's Ample Constitutional Authority to Legislate GHG Emissions through Regulations, a National Cap and Trade Program, or a National Carbon Tax." *National*

Journal of Constitutional Law 36 (331). Open access. https://papers.ssrn.com/sol3/papers.cfm?abstract_id=2775370.

Doelle, Meinhard. 2019. "The Heart of the Paris Rulebook: Communicating NDCs and Accounting for Their Implementation." *Climate Law* 9 (3). Open access, SSRN. https://papers.ssrn.com/sol3/papers.cfm?abstract_id=3332792.

European Commission, Emissions Database for Global Atmospheric Research [EDGAR]. 2020. *Fossil CO_2 Emissions of All World Countries 2020 Report.* https://publications.jrc.ec.europa.eu/repository/handle/JRC121460.

Friedlingstein, P., et al. 2020. "Global Carbon Budget 2020." *Earth System Science Data* 12: 3269–340. Open access. https://doi.org/10.5194/essd-12-3269-2020.

Global Energy Monitor. 2021a. "China Dominates 2020 Coal Plant Development." February. https://globalenergymonitor.org/wp-content/uploads/2021/02/China-Dominates-2020-Coal-Development.pdf.

– 2021b. "Boom and Bust: Tracking the Global Coal Plant Pipeline." April. https://globalenergymonitor.org/wp-content/uploads/2021/04/BoomAndBust_2021_final.pdf.

Hof, A.E., et al. 2017. "Global and Regional Abatement Costs of Nationally Determined Contributions (NDCs) and of Enhanced Action to Levels Well Below 2°C and 1.5°C." *Environmental Science & Policy* 71: 30–40. Open access. https://doi.org/10.1016/j.envsci.2017.02.008.

International Energy Agency. 2021a. *Global Energy Review 2021.* April. https://www.iea.org/reports/global-energy-review-2021.

– 2021c. *Sustainability Recovery Tracker.* October. https://www.iea.org/reports/sustainable-recovery-tracker.

– 2021d. *Coal 2021: Analysis and Forecast to 2024.* December. https://iea.blob.core.windows.net/assets/f1d724d4-a753-4336-9f6e-64679fa23bbf/Coal2021.pdf.

Karlsson-Vinkhuyzen, Sylvia I., et al. 2017. "Entry into Force and Then? The Paris Agreement and State Accountability." *Climate Policy* 18, no. 5: 593–9. Open access. https://www.tandfonline.com/doi/full/10.1080/14693062.2017.1331904.

Leiss, William, Michael Tyshenko, Patricia Larkin, and Daniel Krewski. 2020. "Treaty Framing and Climate Science: Challenges in Managing the Risks of Global Warming." *International Journal of Global Environmental Issues* 19, nos. 1/2/3: 273–93. Open access. https://www.inderscience.com/info/inarticle.php?artid=114882.

Lenton, Timothy, et al. 2019. "Climate Tipping Points – Too Risky to Bet Against." *Nature* 575 (28 November): 592–6. Open access. https://media.

nature.com/original/magazine-assets/d41586-019-03595-0/d41586-019-03595-0.pdf.

Mooney, Chris, et al. 2021. "Countries' Climate Pledges Based on Flawed Data." *Washington Post*, 7 November. https://www.washingtonpost.com/climate-environment/interactive/2021/greenhouse-gas-emissions-pledges-data/?itid=sn_climate%20&%20environment_4/.

NewClimate Institute et al. 2021. *Greenhouse Gas Mitigation Scenarios for Major Emitting Countries*. https://www.pbl.nl/sites/default/files/downloads/pbl-new-climate-institute-iiasa-2021-ghg-mitigation-scenarios-for-major-emitting-countries-2021-update_4527.pdf.

Olivier, J.G.J., and J.A.H.W. Peters. 2020. *Trends in Global CO_2 and Total Greenhouse Gas Emissions*. PBL Netherlands Environmental Assessment Agency, December. https://www.pbl.nl/en/publications/trends-in-global-co2-and-total-greenhouse-gas-emissions-2020-report.

Peters, Glen P., et al. 2015. "Measuring a Fair and Ambitious Climate Agreement Using Cumulative Emissions." *Environmental Research Letters* 10: 105004. Open access. https://iopscience.iop.org/article/10.1088/1748-9326/10/10/105004.

– 2019. "Carbon Dioxide Emissions Continue to Grow Amidst Slowly Emerging Climate Policies." *Nature Climate Change* 10: 3–6. https://doi.org/10.1038/s41558-019-0659-6.

Popovich, Nadia, and Brad Plumer. 2021. "Who Has the Most Historical Responsibility for Climate Change?" *New York Times*, 12 November. https://www.nytimes.com/interactive/2021/11/12/climate/cop26-emissions-compensation.html.

Ritchie, Hannah, and Max Roser. 2021. "CO_2 and Greenhouse Gas Emissions: Annual CO_2 Emissions by World Region." *Our World in Data*. https://ourworldindata.org/co2-and-other-greenhouse-gas-emissions.

United Nations Framework Convention on Climate Change (UNFCCC). 2008. *Kyoto Protocol Reference Manual*. https://unfccc.int/resource/docs/publications/08_unfccc_kp_ref_manual.pdf.

– 2021. "Nationally-Determined Contributions under the Paris Agreement: Synthesis Report by the Secretariat." 17 September. https://unfccc.int/sites/default/files/resource/cma2021_08_adv.pdf.

Vinichenko, V., et al. 2021. "Historical Precedents and Feasibility of Rapid Coal and Gas Decline Required for the 1.5 C Target." *One Earth* 4 (October 22): 1477–90. Open access. https://www.cell.com/one-earth/pdf/S2590-3322(21)00534-0.pdf.

CHAPTER SEVEN

Bank for International Settlements (BIS). 2020. *The Green Swan: Central Banking and Financial Stability in the Age of Climate Change.* https://www.bis.org/publ/othp31.htm.

Bataille, Chris. 2018. "A Review of Technology and Policy Deep Carbonization Pathways for Making Energy-Intensive Industry Production Consistent with the Paris Agreement." *Journal of Cleaner Production* 187: 960–73. Open access. https://epub.wupperinst.org/frontdoor/deliver/index/docId/6984/file/6984_Bataille.pdf.

Belfer Center for Science and International Affairs. 2019. *Governance of the Deployment of Solar Geoengineering.* Open access. https://www.belfercenter.org/publication/governance-deployment-solar-geoengineering.

Bernstein, Steven, and Matthew Hoffmann. 2018. "Decarbonization: The Politics of Transformation." Chapter 14 in *Governing Climate Change*, edited by A. Jordan et al., 248–65. Cambridge: Cambridge University Press. Open access. https://link.springer.com/article/10.1007/s11077-018-9314-8.

BlackRock, Inc. "Our 2021 Stewardship Expectations." https://www.blackrock.com/corporate/literature/fact-sheet/blk-responsible-investment-engprinciples-global.pdf.

Boettcher, Miranda, and Stefan Schäfer, eds. 2017. "Crutzen +10: Reflecting on 10 Years of Geoengineering Research." Special issue. *Earth's Future* 5. https://agupubs.onlinelibrary.wiley.com/doi/toc/10.1002/(ISSN)2328-4277.GEOENGIN1.

Canadian Institute for Climate Choices (CICC). 2020. *Expert Assessment of Climate Pricing Systems.* https://publications.gc.ca/collections/collection_2021/eccc/En4-434-2021-eng.pdf.

Carlarne, Cinnamon. 2014 "Delinking International Law and Climate Change." *Michigan Journal of Environmental and Administrative Law* 4, no. 1: 1–60. Open access. https://repository.law.umich.edu/cgi/viewcontent.cgi?article=1029&context=mjeal.

Coyne, Andrew. 2021a. "A Higher Carbon Price Could Get Us to Paris on Its Own at Much Less Cost to the Economy." *Globe and Mail*, 27 November. https://www.theglobeandmail.com/opinion/article-a-higher-carbon-price-could-get-us-to-paris-on-its-own-at-much-less/.

Deep Decarbonization Pathways Project. 2015a. *Pathways to Deep Decarbonization 2015 Report*, SDSN – IDDRI. https://www.iddri.org/en/publications-and-events/report/pathways-deep-decarbonization-2015-synthesis-report.

– 2015b. *Pathways to Deep Decarbonization in Canada*. https://electricity.ca/wp-content/uploads/2017/05/DDPP_CAN.pdf.

Environment and Climate Change Canada (ECCC). 2018. "Estimated Results of the Federal Carbon Pollution Pricing System." https://www.canada.ca/content/dam/eccc/documents/pdf/reports/estimated-impacts-federal-system/federal-carbon-pollution-pricing-system_en.pdf.

– 2020. *Pan-Canadian Approach to Pricing Carbon Pollution: Interim Report, 2020*. https://publications.gc.ca/collections/collection_2021/eccc/En4-423-1-2021-eng.pdf.

Gertner, Jon. 2021. "Has the Carbontech Revolution Begun?" *New York Times*, 23 June. https://www.nytimes.com/2021/06/23/magazine/interface-carpet-carbon.html.

"Global Climate Action Needs Trusted Financial Data." 2021. *Nature* 589 (7 January). https://media.nature.com/original/magazine-assets/d41586-020-03646-x/d41586-020-03646-x.pdf.

International Energy Agency (IEA). 2021e. *Financing Clean Energy Transitions in Emerging and Developing Economies*. https://www.iea.org/reports/financing-clean-energy-transitions-in-emerging-and-developing-economies.

International Journal of Risk Assessment and Management (IJRAM). 2019. *Risk Assessment and Risk Management of Geologic Storage of Carbon*: Open access. https://www.inderscience.com/info/inarticletoc.php?jcode=ijram&year=2019&vol=22&issue=3/4

Jiang, X., et al. 2018. "Global Rules Mask the Mitigation Challenge Facing Developing Countries." *Earth's Future* 7. Open access. https://agupubs.onlinelibrary.wiley.com/doi/epdf/10.1029/2018EF001078.

Lawrence, Mark G., et al. 2018. "Evaluating Climate Geoengineering Proposals in the Context of the Paris Agreement Temperature Goals." *Nature Communications* 9: 3734. Open access. https://www.nature.com/articles/s41467-018-05938-3.

Leiss, William. 2019. "A Global Decarbonization Bond." *Environmental Research Letters* 14: 091003. Open access. https://iopscience.iop.org/article/10.1088/1748-9326/ab396f.

Lewis, S.L., et al. 2019. "Assessing Contributions of Major Emitters' Paris-Era Decisions to Future Temperature Extremes." *Geophysical Research Letters* 10.1029/2018GL081608: 3936–43. Open access. https://agupubs.onlinelibrary.wiley.com/doi/full/10.1029/2018GL081608.

McEvoy, David M., and Todd L. Cherry. 2016. "The Prospects for Paris: Behavioral Insights into Unconditional Cooperation on Climate Change." *Palgrave Communications* 2: 16056. Open access. https://www.nature.com/articles/palcomms201656.

Martin-Roberts, E., et al. 2021. "Carbon Capture and Storage at the End of a Lost Decade." *One Earth* 4 (19 November): 1–16. Open access. https://www.cell.com/action/showPdf?pii=S2590-3322%2821%2900541-8.

National Academies Press (NAP). 2019. *Negative Emissions Technologies and Reliable Sequestration: A Research Agenda.* http://nap.naptionalacademies.org/download/25259.

Oxfam. 2020. *Climate Finance Shadow Report, 2020.* https://oxfamilibrary.openrepository.com/bitstream/handle/10546/621066/bp-climate-finance-shadow-report-2020-201020-en.pdf.

Renforth, Phil, and Jennifer Wilcox. 2020. *The Role of Negative Emissions Technologies in Addressing our Climate Goals* (ebook), *Frontiers in Climate,* 28 January 2020. https://www.frontiersin.org/research-topics/9752/the-role-of-negative-emission-technologies-in-addressing-our-climate-goals.

Rennert, Kevin, et al. 2021. *The Social Cost of Carbon.* Brookings Papers on Economic Activity, Working Paper (9 September). https://www.brookings.edu/wp-content/uploads/2021/09/Social-Cost-of-Carbon_Conf-Draft.pdf.

Ritchie, Hannah. 2017. "How Much Will It Cost to Mitigate Climate Change?" *Our World in Data* (blog post), 27 May. https://ourworldindata.org/how-much-will-it-cost-to-mitigate-climate-change.

Rivers, Nicholas, and Randall Wigle. 2018. "Reducing Greenhouse Gas Emissions in Transport: All in One Basket?" *School of Public Policy Publications* 11, no. 5 (February). https://papers.ssrn.com/sol3/papers.cfm?abstract_id=3116331.

"Statement by Eighteen Countries at Paris in September 2015." https://unfccc.int/news/18-industrial-states-release-climate-finance-statement.

United Nations (UN). 2020. *Delivering on the $100 Billion Climate Finance Commitment and Transforming Climate Finance.* December. https://www.un.org/sites/un2.un.org/files/climate_finance_report.pdf.

– 2021. *Climate Finance Delivery Plan.* 25 October. https://ukcop26.org/wp-content/uploads/2021/10/Climate-Finance-Delivery-Plan-1.pdf.

United Nations Framework Convention on Climate Change (UNFCCC). 2018. Talanoa Dialogue "Call for Action." https://unfccc.int/sites/default/files/resource/Talanoa%20Call%20for%20Action.pdf.

United States, National Academies Press. 2021. *Accelerating Decarbonization of the U.S. Energy System.* https://www.nap.edu/download/25932.

– 2021. *Reflecting Sunlight: Recommendations for Solar Geoengineering Research and Research Governance.* https://www.nap.edu/catalog/25762/reflecting-sunlight-recommendations-for-solar-geoengineering-research-and-research-governance.

Victor, David G., et al. 2009. "The Geoengineering Option." *Foreign Affairs*, March/April. Open access. https://fsi-live.s3.us-west-1.amazonaws.com/s3fs-public/The_Geoengineering_Option.pdf.

Wagner, G., et al. 2021. "Eight Priorities for Counting the Social Cost of Carbon." *Nature* 590. Open access. https://media.nature.com/original/magazine-assets/d41586-021-00441-0/d41586-021-00441-0.pdf.

World Resources Institute (WRI). 2021a. *A Breakdown of Developed Countries' Public Climate Finance Contributions towards the $100 Billion Goal.* October. https://files.wri.org/d8/s3fs-public/2021-10/breakdown-developed-countries-public-climate-finance-contributions-towards-100-billion.pdf.

Topics and Recommended Websites

"CO$_2$ and Greenhouse Gas Emissions" (Our World in Data).

"Generation IV Nuclear Reactor."

"Molten Salt Reactor" (Terrestrial Energy [Canada]).

"Stratospheric Aerosol Injection" (Wikipedia).

Carbonengineering.com.

CHAPTER EIGHT

MITIGATION

Auditor General of Canada. 2021. *Lessons Learned from Canada's Record on Climate Change.* Report of the Commissioner of the Environment and Sustainable Development. https://www.oag-bvg.gc.ca/internet/English/parl_cesd_202111_05_e_43898.html.

– 2022. *2022 Reports 1 to 5 of the Commissioner of the Environment and Sustainable Development.* https://www.oag-bvg.gc.ca/internet/English/parl_cesd_202204_e_44020.html.

Barecka, M., et al. 2021. "Carbon Neutral Manufacturing via On-Site CO2 Recycling." *iScience* 24 (June 25): 102514. Open access. https://www.cell.com/iscience/pdf/S2589-0042(21)00482-X.pdf.

Beugin, Dale. 2020. "Canada's New Climate Plan Is a Big Deal – Here's Why." Canadian Institute for Climate Choices. 11 December. https://climate-choices.ca/canadas-climate-plan/.

Boyce, Mark S. 2021. "Mimic the Bison: Why We Should Bury Carbon Tax Revenues in Soil." *Globe and Mail*, 5 May. https://www.theglobeandmail.com/canada/article-mimic-the-bison-why-we-should-bury-carbon-tax-revenues-in-soil/.

Canada. 2021. "Canada's 2021 Nationally Determined Contribution under the Paris Agreement." https://www4.unfccc.int/sites/ndcstaging/

PublishedDocuments/Canada%20First/Canada's%20Enhanced%20NDC%20
Submission1_FINAL%20EN.pdf.

Canada Energy Regulator (CER). 2020. *Canada's Energy Future 2020.* https://
www.cer-rec.gc.ca/en/data-analysis/canada-energy-future/2020/canada-
energy-futures-2020.pdf.

– 2021. *Canada's Energy Future 2021.* https://www.cer-rec.gc.ca/en/data-
analysis/canada-energy-future/2021/canada-energy-futures-2021.pdf.

Canada. Environment and Climate Change Canada. 2020. *Canada's National
Report on Black Carbon and Methane.* http://publications.gc.ca/collections/
collection_2021/eccc/En11-18-2021-eng.pdf.

– 2020. *A Healthy Environment and a Healthy Economy.* https://www.canada.
ca/en/services/environment/weather/climatechange/climate-plan/
climate-plan-overview/healthy-environment-healthy-economy.html.

– 2020. *Modelling and Analysis of "A Healthy Environment and a Healthy
Economy."* https://www.canada.ca/content/dam/eccc/documents/pdf/
climate-change/climate-plan/annex_modelling_analysis_healthy_
environment_healthy_economy.pdf.

– 2020. *National Inventory Report, 1990–2018: Greenhouse Gas Sources and
Sinks in Canada,* Parts 1, 2, 3. https://unfccc.int/documents/224829.

– 2020. *Progress towards Canada's Greenhouse Gas Emissions Reduction Target.*
https://www.canada.ca/content/dam/eccc/documents/pdf/cesindicators/
progress-towards-canada-greenhouse-gas-reduction-target/2020/
progress-ghg-emissions-reduction-target.pdf.

– 2021. *Canada's Greenhouse Gas and Air Pollutant Emissions Projections 2020.*
http://publications.gc.ca/collections/collection_2021/eccc/En1-78-2020-
eng.pdf.

– 2021. *Progress towards Canada's Greenhouse Gas Emissions Reduction Target.*
https://www.canada.ca/content/dam/eccc/documents/pdf/cesindicators/
progress-towards-canada-greenhouse-gas-reduction-target/2021/
progress-ghg-emissions-reduction-target.pdf.

– 2022. *National Inventory Report 1990–2019: Greenhouse Gas Sources and
Sinks in Canada, Executive Summary 2022.* https://www.canada.ca/
en/environment-climate-change/services/climate-change/
greenhouse-gas-emissions/sources-sinks-executive-summary-
2022.html.

Canada. Farmers for Climate Solutions. 2021. "A Down Payment for a Resilient
Farm Future: Budget 2021 Recommendation." https://farmersforclimate
solutions.ca/budget-2021-recommendation/#programs.

Canada. Net-Zero Advisory Body. 2021. "Net-Zero Pathways: Initial
Observations." June. https://nzab2050.ca/publications.

Clean Prosperity. 2021. *Assessing the 2021 Federal Liberal Climate Plan*. https://cleanprosperity.ca/wp-content/uploads/2021/10/Clean_Prosperity_LPC_Climate_Policy_Report_2021.pdf.

Climate Action Tracker. "NDC Ratings and LULUCF." Accessed 6 June 2022. https://climateactiontracker.org/methodology/indc-ratings-and-lulucf/.

Coyne, Andrew. 2021b. "Is Carbon Pricing Liberal Policy?" *Globe and Mail*, 5 November. https://www.theglobeandmail.com/opinion/article-is-carbon-pricing-liberal-policy-for-the-most-part-its-anything-but/.

Drever, C. Ronnie, et al. 2021. "Natural Climate Solutions for Canada." *Science Advances 7*. Open access. https://advances.sciencemag.org/content/advances/7/23/eabd6034.full.pdf.

Emissions Database for Global Atmospheric Research (EDGAR). European Commission. 2021. *GHG Emissions of All World Countries 2021 Report*. https://edgar.jrc.ec.europa.eu/report_2021.

Environment and Climate Change Canada (ECCC). 2022. *Canada's 2030 Emissions Reduction Plan*. https://www.canada.ca/content/dam/eccc/documents/pdf/climate-change/erp/Canada-2030-Emissions-Reduction-Plan-eng.pdf.

Friedlingstein, P., et al. 2021. "Global Carbon Budget 2021." *Earth System Science Data*. Open access. https://doi.org/10.5194/essd-2021-386.

Fyson, C.L., and M.L. Jeffery. 2019. "Ambiguity in the Land Use Component of Mitigation Contributions toward the Paris Agreement Goals." *Earth's Future* 7: 873–91. Open access. https://agupubs.onlinelibrary.wiley.com/doi/pdf/10.1029/2019EF001190.

Gelles, David. 2022. "A Fight over America's Energy Future Erupts on the Canadian Border." *New York Times*, 6 May. https://www.nytimes.com/2022/05/06/climate/hydro-quebec-maine-clean-energy.html.

Government of Alberta. 2021. "Public Inquiry into Anti-Alberta Energy Campaigns." Comment by Martin Olszynski. http://ablawg.ca/wp-content/uploads/2021/01/Blog_MO_Public_Inquiry_AAEC.pdf.

Harris, L.I., et al. 2021. "The Essential Carbon Service Provided by Northern Peatlands." *Frontiers in Ecology and the Environment* 20, no. 4: 222–30. Open access. https://esajournals.onlinelibrary.wiley.com/doi/epdf/10.1002/fee.2437.

Hughes, Larry. 2021. "How Canada Intends to Achieve Its 2030 Emissions Targets." *Policy Options*, July 2021. https://policyoptions.irpp.org/magazines/july-2021/how-canada-intends-to-achieve-its-2030-emissions-targets/.

Jaccard, Mark, and Bradford Griffin. 2021. *A Zero-Emission Canadian Electricity System by 2035*. David Suzuki Foundation. August. https://

davidsuzuki.org/wp-content/uploads/2021/08/Jaccard-Griffin-Zero-emission-electricity-DSF-2021.pdf.

Jackson, R.B., et al. 2021. "Global Fossil Carbon Emissions Rebound Near Pre-COVID-19 Levels." *Environmental Research Letters* 17, no. 3. Open access. https://iopscience.iop.org/article/10.1088/1748-9326/ac55b6.

McClearn, Matthew. 2021. "Canada's First New Nuclear Reactor in Decades Is an American Design." *Globe and Mail*, 26 December. https://www.theglobeandmail.com/business/article-canadas-first-new-nuclear-reactor-in-decades-is-an-american-design/.

Miller, S.A., et al. 2021. "Achieving Net Zero Greenhouse Gas Emissions in the Cement Industry via Value Chain Mitigation Strategies." *One Earth* 4, no. 9 (October 22): 1398–1411. Open access. https://www.cell.com/one-earth/pdf/S2590-3322(21)00533-9.pdf.

Plumer, B., and N. Popovich. 2021. "Yes, There Has Been Progress on Climate. No, It's Not Nearly Enough." *New York Times*, 25 October. https://www.nytimes.com/interactive/2021/10/25/climate/world-climate-pledges-cop26.html.

Princeton University. 2021. *Net-Zero America: Final Report Summary.* 29 October. https://acee.princeton.edu/rapidswitch/projects/net-zero-america-project/.

Reguly, Eric. 2021. "The Government's 2035 Electrical Vehicle Mandate Is Delusional." *Globe and Mail*, 3 July. https://www.theglobeandmail.com/business/commentary/article-the-governments-2035-electric-vehicle-mandate-is-delusional/.

Rissman, Jeffrey, et al. 2020. "Technologies and Policies to Decarbonize Global Industry: Review and Assessment of Mitigation Drivers through 2070." *Applied Energy* 266: 114848. Open access. https://www.sciencedirect.com/science/article/pii/S0306261920303603.

– Supplementary Material (graphics). https://ars.els-cdn.com/content/image/1-s2.0-S0306261920303603-fx1_lrg.jpg.

Royal Bank of Canada (RBC). 2021. *The $2 Trillion Transition: Canada's Road to Net-Zero.* October. https://royal-bank-of-canada-2124.docs.contently.com/v/the-2-trillion-transition-canadas-road-to-net-zero-pdf.

SEI (SEI, IISD, ODI, E3G, and UNEP). 2021. *The Production Gap Report 2021.* http://productiongap.org/2021report.

Semeniuk, Ivan. 2021b. "What Lies Beneath: Exploring Canada's Invisible Carbon Storehouse." *Globe and Mail*, 10 November. https://www.theglobeandmail.com/canada/article-what-lies-beneath-exploring-canadas-invisible-carbon-storehouse/.

Smil, Vaclav. 2022. "This Eminent Scientists Says Climate Activists Need to Get Real." Interview by David Marchese. *New York Times Magazine*, 25 April 2022. https://www.nytimes.com/interactive/2022/04/25/magazine/vaclav-smil-interview.html.

Sothe, C., et al. 2021. "Large Soil Carbon Storage in Terrestrial Systems in Canada." *Global Biogeological Cycles*. Open access. https://doi.org/10.1002/essoar.10507117.2.

Sothe, C., et al. 2022. "Large Scale Mapping of Soil Organic Carbon Concentration with 3D Machine Learning and Satellite Observations." *Geoderma* 405: 115402. Open access. https://doi.org/10.1016/j.geoderma.2021.115402.

Trottier Energy Institute/Institut de l'énergie Trottier. 2021. *Canadian Energy Outlook 2021: Horizon 2060* (October). https://iet.polymtl.ca/wp-content/uploads/delightfuldownloads/CanadianEnergyOutlook2021.pdf.

United Nations Environment Programme. 2020. *Emissions Gap Report 2020*. https://www.unenvironment.org/emissions-gap-report-2020.

– 2021. *Emissions Gap Report 2021*. https://www.unep.org/resources/emissions-gap-report-2021.

United States. National Oceanic and Atmospheric Administration (NOAA). 2022. "Increase in Atmospheric Methane Set Another Record during 2021." 7 April. https://www.noaa.gov/news-release/increase-in-atmospheric-methane-set-another-record-during-2021.

Vogl, V., et al. 2021. "Phasing Out the Blast Furnace to Meet Global Climate Targets." *Joule* 5 (October 20): 2646–62. Open access. https://www.cell.com/joule/pdf/S2542-4351(21)00435-9.pdf.

Wesseling, J.H., et al. 2017. "The Transition of Energy Intensive Processing Industries towards Deep Decarbonization: Characteristics and Implications for Future Research." *Renewable and Sustainable Energy Reviews* 79: 1301–13. Open access. https://www.sciencedirect.com/science/article/pii/S1364032117307906.

Wildlife Conservation Society Canada (WCSC). 2021. "Northern Peatlands in Canada: An Enormous Carbon Storehouse." https://storymaps.arcgis.com/stories/19d24f59487b46f6a011dba140eddbe7.

World Meteorological Organization (WMO). 2021. *WMO Greenhouse Gas Bulletin*, No. 17 (25 October). https://reliefweb.int/report/world/wmo-greenhouse-gas-bulletin-state-greenhouse-gases-atmosphere-based-global-2.

World Resources Institute (WRI). 2021. *State of Climate Action 2021: Systems Transformations Required to Limit Global Warming to 1.5°*. https://www.wri.org/research/state-climate-action-2021.

Topics and Recommended Websites:

https://www.terrestrialenergy.com/.

IMPACTS AND ADAPTATION

Arctic Institute. 2021. "Climate Change and Geopolitics: Monitoring of a Thawing Permafrost." https://www.thearcticinstitute.org/climate-change-geopolitics-monitoring-thawing-permafrost/.

Berkeley Earth. 2021. "Actionable Climate Science for Policymakers: Country-Level Warming Projections." http://berkeleyearth.org/policy-insights/?mc_cid=b99a9b467f&mc_eid=43ca8fffe8.

Burke, Marshall, et al. 2015. "Global Non-linear Effect of Temperature on Economic Production." *Nature* 527: 235–9. https://doi.org/10.1038/nature15725. Available at http://emiguel.econ.berkeley.edu/assets/miguel_research/66/BurkeHsiangMiguel2015.pdf.

– 2015. "Economic Impact of Climate Change on the World." Open access. https://web.stanford.edu/~mburke/climate/map.php.

Canada. 2022. Environment and Climate Change Canada. *Canada's Changing Climate Report 2022.* https://ftp.maps.canada.ca/pub/nrcan_rncan/publications/STPublications_PublicationsST/329/329703/gid_329703.pdf.

Canada. Library of Parliament. 2020. *Climate Change: Its Impacts and Policy Implications.* https://lop.parl.ca/staticfiles/PublicWebsite/Home/ResearchPublications/BackgroundPapers/PDF/2019-46-e.pdf.

Diffenbaugh, Noah S., and Marshall Burke. 2019. "Global Warming Has Increased Global Economic Inequality." *Proceedings of the National Academy of Sciences* 116, no. 20 (May): 9808–13. Open access. https://www.pnas.org/content/116/20/9808.

Heslin, Alison, et al. 2020. "Simulating the Cascading Effects of an Extreme Agricultural Production Shock: Global Implications of a Contemporary US Dust Bowl Event." *Frontiers in Sustainable Food Systems*, 20 March. Open access. https://www.frontiersin.org/articles/10.3389/fsufs.2020.00026/full.

Intergovernmental Panel on Climate Change (IPCC). 2014. "Summary for Policymakers." In *Climate Change 2014: Impacts, Adaptation and Vulnerability.* Cambridge: Cambridge University Press. https://www.ipcc.ch/site/assets/uploads/2018/02/ar5_wgII_spm_en.pdf.

Lustgarten, Abrahm. 2020. "How Russia Wins the Climate Crisis." *New York Times Magazine*, 16 December. https://www.nytimes.com/interactive/2020/12/16/magazine/russia-climate-migration-crisis.html.

National Oceanic and Atmospheric Administration (NOAA). 2020. *Arctic Report Card 2020*. https://arctic.noaa.gov/Portals/7/ArcticReportCard/Documents/ArcticReportCard_full_report2020.pdf.

– 2021. *Arctic Report Card 2021*. https://arctic.noaa.gov/Portals/7/ArcticReportCard/Documents/ArcticReportCard_full_report2021.pdf.

Parfenova, Elena, et al. 2020. "Assessing Landscape Potential for Human Sustainability and 'Attractiveness' across Asian Russia in a Warmer 21st Century." *Environmental Research Letters* 14: 065004. Open access. https://iopscience.iop.org/article/10.1088/1748-9326/ab10a8/pdf.

Pisor, Anne C., and James H. Jones. 2020. "Human Adaptation to Climate Change." *American Journal of Human Biology* 33, no. 4. Open access. https://onlinelibrary.wiley.com/doi/epdf/10.1002/ajhb.23530.

Ritchie, Hannah. 2021. "Who Has Contributed Most to Global CO_2 Emissions?" *Our World in Data*. https://ourworldindata.org/contributed-most-global-co2.

Ritchie, Hannah, and Max Roser. 2021. "Canada's CO_2 Country Profile." *Our World in Data*. https://ourworldindata.org/co2/country/canada?country=~CAN.

Rodriguez-Fernández, Laura, et al. 2020. "Allocation of Greenhouse Gas Emissions Using the Fairness Principle: A Multi-country Analysis." *Sustainability* 12 (14): 5839. Open access. https://www.mdpi.com/2071-1050/12/14/5839.

Sweet, W.V., et al. 2022. *Global and Regional Sea Level Rise Scenarios for the United States: Updated Mean Projections and Extreme Water Level Probabilities along U.S. Coastlines*. US National Oceanic and Atmospheric Administration. https://aambpublicoceanservice.blob.core.windows.net/oceanserviceprod/hazards/sealevelrise/noaa-nos-techrpt01-global-regional-SLR-scenarios-US.pdf.

United Nations Framework Convention on Climate Change (UNFCCC). 2021. *Nationally-Determined Contributions under the Paris Agreement: Synthesis Report by the Secretariat*. 17 September. https://unfccc.int/sites/default/files/resource/cma2021_08_adv.pdf.

Topics and Recommended Websites

"Climate Change Mitigation" (Wikipedia).
"Climate Change Adaptation" (Wikipedia).

REMINISCENCES AND ACKNOWLEDGMENTS

Warner, Gerry. 2018. "Time to End Climate Change Debate Says Oil Exec." Op-ed, 9 June. *e-know.ca*. https://www.e-know.ca/regions/east-kootenay/time-to-end-climate-change-debate-says-oil-exec/.

APPENDIX ONE

Bloomberg News. 2021. "The Chinese Companies Polluting the World More Than Entire Nations." 24 October. https://www.bloomberg.com/graphics/2021-china-climate-change-biggest-carbon-polluters/.

Rhodium Group. 2021. "China's Greenhouse Gas Emissions Exceeded the Developed World for the First Time in 2019." 6 May. https://rhg.com/research/chinas-emissions-surpass-developed-countries/. See also https://rhg.com/research/preliminary-2020-global-greenhouse-gas-emissions-estimates/.

Index

Index

Index